AD 151
EONK
2-S

THE ——————
WASTED
OCEAN

DATE DUE

THE WASTED OCEAN

DAVID K. BULLOCH

Lyons & Burford, Publishers

AN AMERICAN LITTORAL SOCIETY BOOK

#32

Copyright © 1989 by David K. Bulloch.

ALL RIGHTS RESERVED. No part of this book may be
reproduced in any manner without the express written
consent of the publisher, except in the case of brief
excerpts in critical reviews and articles. All inquiries should
be addressed to: Lyons & Burford, 31 West 21 Street,
New York, NY 10010.

Printed in the United States of America

10 9 8 7 6 5 4 3 2 1

Library of Congress Cataloging-in-Publication Data

Bulloch, David K.
    The wasted ocean / David K. Bulloch.
        p.   cm.
    Bibliography: p.
    Includes index.
    ISBN 1-55821-019-9 : $16.95.—ISBN 1-55821-034-2 (pbk.) : $9.95
    1. Marine pollution—Environmental aspects.   I. Title.
QH545.W3B85   1989
363.7′394′09162—dc19

                                                        89-30025
                                                          CIP

CONTENTS

About the American Littoral Society

The American Littoral Society is a non-profit environmental organization dedicated to the study and conservation of the littoral zone—those fragile, fertile areas where the sea meets the land. Marine life concentrates here. Plankton, shellfish, crab, fish, waterfowl, and shorebird productivity depends on the availability and health of these special places. Increasingly man concentrates here too. To learn more about how we can coexist with the ocean's living bounty, please write the Society for information about its programs, publications, and activities: AMERICAN LITTORAL SOCIETY, HIGHLANDS, NJ 07732. (201) 291-0055.

ACKNOWLEDGEMENTS

Gathering the myriad facts and winnowing out the vital concepts for a topic as big as this one was considerably eased by the aid of a number of people. C.D. Hardy of Southampton College offered his reprint collection. Dr. John Pearce of the National Marine Fisheries Service (NMFS) at Woods Hole supplied much specific data on eastern estuaries. Claire Steimle and Judy Berrien helped me mine information from the NMFS library at Sandy Hook, NJ.

Conversations with Frank Steimle and Tony Pacheco, both with NMFS, Beth Millemann of the Coast Alliance, Todd Miller of the North Carolina Coastal Federation, Don Riepe of the National Park Service, Dave Grant of Brookdale College, Pearl Schwartz of the League of Women Voters, Trudy Coxe of Save The Bay, R.I., Cindy Zipf of Clean Ocean Action and Dery Bennett of the American Littoral Society cleared up a multitude of questions. Lou and Helen Burlingame provided the drawings.

And, of course, Edie, who labored uncomplaining with manuscript and all the other desiderata that months of scribbling engenders.

INTRODUCTION

A wise man once suggested that our business and political leaders should ask themselves every day, how many people can the earth support, and at what standard of living?

So far, few businessmen or politicians anywhere have grasped the importance—nay, essentiality—of that combination question. The awareness that mankind has grossly overloaded the circuitry of our global ecosystem is growing less fast than our ever-swelling demand for food, housing, clothing, and recreation. The Population Reference Bureau estimates that the net increase of the world's population is 10,185 people *per hour*. This number represents individuals who must be fed, housed, clothed, employed and, most important of all, educated, if they are to achieve something more than mere existence.

We in North America are fortunate in having more fresh water, more fertile soils, and generally more resources than any other continent. Indeed, nature's fecundity enabled democracy to take root and flourish here in the first place. Yet not even this earthly paradise can sustain limitless numbers of people, especially when our business and political leaders have allowed so much of our water to be poisoned and so many of our resources to be wasted. Nowhere is our wantonness more apparent than in that most fragile and essential realm, the interface of sea and shore known as the littoral zone.

Our precariousness has made environmentalists of countless thousands of people who formerly didn't know a periwinkle from a poppy. Some individuals have accomplished much, but many have accomplished more as members of specialized non-governmental conservation organizations. In the past, their challenge was to create sufficient environmental awareness and legislation to safeguard threatened ecosystems. The sustaining challenge of tomorrow will be to educate and persuade local wetland boards, zoning commissions, and "biocrats" in all the coastal and near-coastal counties, boroughs, townships and cities of North America to uphold the law and to regulate the exploitation of marine resources with restraint and common sense.

Dave Bulloch has provided us with the first handbook of marine conservation. He is an industrial chemist by vocation, and a citizen-activist by avocation. Technically trained and better aware than most people of the insidious side-effects of many man-made compounds and residues, David has also been a member and officer of the American Littoral Society almost from its inception in 1962.

This book is not a mere doomsday account of the ocean's ills. Bulloch is well aware that it's no longer enough to publicize problems; the greatest need is to find ways to involve the public in solving those problems. In the past, the majority have been reluctant to participate, because they were overwhelmed by the magnitude and diversity of marine conservation crises, ranging from thoughtless development onshore to the dumping of toxic wastes offshore. People get involved only when they realize they can make a difference. This book describes not only most every aspect of oceanic pollution, it describes ordinary people who've made a difference. It suggests ways we can all get involved—indeed, *must* get involved—since state and federal environmental agencies are increasingly more concerned with protecting their own turfs and comfort than with upholding laws designed to protect the public from industrial and municipal abuse.

This book reminds us of the scientific truth that there is no action without reaction. Although some reactions may take a while to become apparent, anything that impacts or alters an estuary's chemistry also ultimately impacts or alters the earth's chemistry as well. We must end our business-as-usual dealings with the sea; we must stop contaminating this most vital of all ecosystems. This book encourages us to believe the job can be done, and shows us the way to do it.

George Reiger,
author of *Wanderer on My Native Shore*

ONE

THE WASTED OCEAN

The coastlines of the United States are under assault. Gulfs and bays, the waters washing our beaches, and the brackish waters flowing through thousands of miles of estuaries, flats, tidelands, and marshes are inexorably being degraded. No longer do they sustain the thriving and diverse stocks of waterfowl, fish, and shellfish of past years. Beaches and shorelines around industrial areas are fouled with oil, awash with sewage, and littered with garbage.

Critical coastal habitats are disappearing. More than one-half of all our marine wetlands have been destroyed—filled in for ports, marinas, or housing or used as dumps. Some have been choked by spoil from channel dredging. Others have eroded away as dams, dikes, and embankments have diverted upriver fresh water or cut off seawater circulation.

Whole ecosystems are collapsing under a bath of excess nutrients and strange new substances. This chemical bombardment flows from countless pipelines, leaches from dumps and fills, and runs off from streets and farms.

Vanishing uplands, now converted into developments or

cleared for agriculture, once held back the rain, absorbing it, filtering it, and slowly releasing it into the streams and rivers that make their way to the wetlands of the nearby coast. These rains now plunge uninhibited into waterways, carrying with them silt, fertilizers, pesticides, and bacteria from farms and oils, heavy metals, particulates, and trash from city streets.

Along our entire coastline, municipal and industrial waste is pumped either into a convenient estuary or directly into the sea. Many municipal systems are combined with street runoff; during heavy rainfalls these raw waste streams are routinely diverted from the sewage plant directly into coastal or estuarine water with no treatment at all.

This wasting of our waters coincides with the increasing popularity of the pleasures of our coast and its seafood. Per-capita consumption of fish and shellfish is at an all-time high. So too are visits to the beach, recreational fishing, boating, and the demand for nearby living space.

The slow collapse of these marine ecosystems draws little attention. It takes a visible catastrophe for degradation to make the news and, when it does, the story fades quickly. Recurring anoxic bottom water off the Louisiana coast merits barely an inch of type in the national press. A fish kill may rate a picture, especially if the rotting carcasses wash ashore on a popular beach, but little is said about why it happened. Trash on the beach rarely makes the papers. To hit the headlines there must be a twist to the story. In summer 1987, a fifty-mile-long garbage slick from careless barge unloading of New York City trash at their Staten Island Fresh Kills landfill snaked southward along the New Jersey shore. Nothing unusual in that, but this time what washed up included medical waste, the discards of New York clinics. Hypodermic needles and syringes were beached ashore and the media had a field day. Tourists shied away, merchants and realtors saw business sag, and the governor of New Jersey raised hell. The season ended and the flurry died down and all but vanished from the news until the following summer. More medical waste washed up, shutting some New York and New Jersey beaches for three to four weeks during one of the hottest summers on record. Jersey shore business collapsed, down 50 percent by early estimates.

Meanwhile, the relentless decimation going on in the once productive backwaters went unheralded, except for a few routine notices about new shellfish-bed closings.

Fortunately many environmental organizations are fighting and, in some areas, winning the battle for the attention and concern of the public over coastal degradation. Some coastal states, whose economies depend heavily on the integrity of their beaches and fisheries, have moved to slow down the wholesale despoliation of these vital areas.

Is it too little, too late? Since 1972 federal legislation has been in force to help stem the destructive tide. Among these laws, the Clean Water Act and the Marine Protection, Research, and Sanctuaries Act regulate the discharge and dumping of wastes into marine waters, put limits on certain kinds and amounts of pollutants, and set standards for sewage treatment.

Although straightforward in concept, this legislation has been difficult to implement. Federal funding for enforcement and monitoring has fallen short, and state and federal money for upgrading municipal sewage systems has dwindled.

The reluctant conclusions of a study by the Office of Technology Assessment are: "Current programs do not adequately address toxic pollutants or non-point source pollution. . . . Pipeline discharge and non-point source pollution (particularly urban runoff) will increase. . . . Federal resources available for municipal sewage treatment are declining . . . and the ability of states or communities to fill the breach is highly uncertain and . . . [many] factors will make it difficult or impossible to shift disposal of certain wastes out of estuaries and coastal waters."

The waste deluge is only one of the impacts on these waters. The demand for land and easy access to these fragile ecosystems grows every year and is destroying the very basis of its fecundity.

The root of the problem lies in the steady increase in nearshore population and shifts in our life-styles and our technology choices.

The United States population is now just over 241 million people and is shifting from the center of the country to its coastlines. From 1950 to 1984, the number of people living in marine coastal counties increased by over 30 million, a rise of more than 80 percent. Right now over 40 percent of our population lives within fifty miles of saltwater coastline. Long stretches of coastline are now a contiguous ribbon of houses, roads, and storefronts.

This problem has been aggravated by federal legislation. As homes were struck by coastal flooding, the federal government in 1968 passed the National Flood Insurance Program to provide protection from water and wind damage in those locations that private insurance carriers would not cover. In 1973, the Flood Disaster Protection Act was passed to assist communities that set up "adequate safeguards and land use restrictions."

It hasn't worked. These laws became an open invitation to build in totally unsuitable areas. The upshot of its effects, aside from saddling the taxpayer with enormous bills when heavy coastal storms hit (which they do with unsurprising regularity) is to continue to encourage new construction in flood zones and promote rebuilding in areas where real estate is repeatedly destroyed.

Concurrent with the population shift that started shortly

after World War II, life-styles and consumption patterns of Americans began to change.

Each U.S. citizen now disposes of three and a half pounds of garbage each day, twice the amount of waste generated by people in any other western country. This throw-away mentality exacerbates local garbage disposal, burdening solid and liquid waste with toxins and confronting the eye with ever more unpleasant reminders of our folly.

Our technological choices have created long-term ills. High-compression auto engines create new sources of nitrogen oxides, causing smog and acid rain. We have chosen chemical fertilizers over manure, pesticides over biological controls, and new nonbiodegradable products over older ones made from natural materials.

What do we stand to lose by the demise of our coastal zones? Among other things, an important source of food, and a great deal of money.

More than two-thirds of our food fish spend a part of their life cycle in estuarine waters. U.S. commercial fishermen caught just over six billion pounds of fish in 1985, the bulk of it taken in coastal waters. Our consumption of fish and shellfish has reached 14.5 pounds per person per year, the highest in our history. But much of it is imported. As our stocks decline, so does the number of fishermen seeking them out. Forced to go farther from home port and fish longer for a reduced catch, many have called it quits. Others who have depended on more than one fishery to see them through the year have been unable to sustain themselves because specific stocks have dwindled. Communities built on fishing are disappearing. Like the family farmer, the independent commercial fisherman is an endangered species.

Besides the commercial fishery, sportfishing is an enormous business. More than 12 million people fished our seacoasts in 1980 and, in doing so, spent $2.4 billion on food, lodging, transportation, and equipment.

New Jersey claims that its shore-centered tourist business is worth $13 billion a year. Florida says 13 million people used their beaches in 1984 and directly and indirectly generated 180,000 jobs. Probably every coastal state has at least a $2 billion a year stake in their shoreline from tourism. Add it all up and you have a stake that's worth keeping.

Above all, the enjoyment and solace that fresh, clean beaches can bring to humans outweighs their direct economic benefits. Whether for swimming, sunning, walking, or just lazy contemplation, coastal areas are priceless. National parklands along marine waters drew 60 million visitors in 1985. Wildlife alone attracted 5 million people to the coasts in 1980.

What is going wrong in these waters? There are as many answers as there are waterways. Each has been burdened with its own mixture of pollutants, and the life within has responded to these insults according to its own kind.

The earliest example of the destruction of a fine food fish was the demise of the Atlantic salmon. Salmon once ran up every seaward-bound river system from Labrador to Connecticut. They were marvelously abundant in colonial days, so inexpensive and plentiful that apprentices in Connecticut balked at eating them more often than twice a week. By 1820 they had dwindled to a trickle there. The last salmon run in the Merrimac River was 1860. By the 1880s only six rivers in Maine produced salmon; by 1925, only two. The salmon's homing instincts brought them fruitlessly back to streams fouled with everything from sawdust to tannery waste until they perished.

The salmon story is old hat. More recent victims are falling like dominoes. The striped bass fishery has collapsed. The Chesapeake Bay oyster is failing and the Peconic Bay scallop is gone. Forty-two percent of all shellfish beds in the

United States are permanently closed. Only 1 percent of California's beds remain open. Almost all of Louisiana's beds are polluted enough to be closed part of the year and are opened only "conditionally" when pollution levels are low enough to allow it.

The canvasback duck population of the Chesapeake Bay has fallen by a factor of eight since 1954. Pollutants have drastically reduced the bay's sea grasses, especially wild celery, the ducks' staple food. No part of the U.S. coastline has been immune from wildlife losses. The litany is endless.

The consequences of this deluge of contaminants and pathogens hits hardest in the semienclosed, low salinity backwaters behind barrier beaches, in embayments, and in the shallows near river mouths where dilution and exchange with the coastal or open ocean waters is slow. Laden with nitrogen- and phosphorus-bearing material from pollution, these static waters may suddenly experience voluminous blooms of algae that swamp out indigenous life. As quickly as this algae bloom explodes, it begins to die. As it decomposes, it uses up the dissolved oxygen in the surrounding water. This process is called eutrophication and can be the death knell for the remaining animals living there. Organic matter, a major component of municipal waste, also consumes oxygen.

All animal life depends on oxygen. Marine animals take theirs from the dissolved oxygen in the water around them. In temperate zones where the winter waters are cold and the summer and fall waters warm, the oxygen content of the water changes with the seasons. Warm water holds less oxygen than does cold.

Bacteria grow faster in warm than in cold water. The two effects, the seasonal increase in temperature and the speeded up production and decay of life, can combine to lower oxygen to a point where animal life not only is threatened but perishes altogether. New bacteria continue the processes of decay,

7

decomposing anaerobically, without the aid of oxygen. This process produces hydrogen sulfide, as poisonous to aquatic life as hydrogen cyanide is to terrestrial life. Except for the bacteria producing it, sulfide-laden waters are lifeless.

Natural clays and silts, washed out of disturbed soil, use up little oxygen but make the water turbid, as do the suspended solids in municipal and industrial waste. Finely divided solids cut off the sunshine that is as vital for the growth of sea grasses as it is for plants on land. Without light, the grasses perish. Vegetation in water holds sediments in place. As sea grasses vanish, the bottom is easily stirred up by storms or fast-moving runoff. Incoming sediments stay suspended, further adding to the cloudiness of the water. Farther down the bay, where the flow slows, settling particulates can foul surfaces and render them uninhabitable for sessile life. Oyster spat, for instance, must find a hard, clean surface before they will attach and begin their transformation into adults. Silty spawning beds quickly eliminate them.

Toxic chemicals harm aquatic life in a variety of ways: some inhibit growth, some interfere with reproduction, some interrupt cell processes—causing death or deformity—and some are absorbed, harmless to their initial host. When that animal is eaten by another higher up on the food chain, the toxins in it can either further accumulate (or biomagnify, as the process is called) or directly damage their new recipient.

Once a synthetic toxin becomes widespread in an ecosystem, it takes a considerable time before it is expunged. Most synthetic compounds have no counterparts in nature, so nature is seldom prepared to break them down. Natural products, even the most toxic substances, meet their match in bacteria that break them down, rendering them into useful material for another generation. But some synthetics, benign or harmful, are so refractory that nothing in nature will decompose them. Plastics are a prime example. Of the billions

8

of pounds of plastics that have been produced, only that which has been burned has recycled. The rest all sits somewhere, unaffected by its surroundings.

A classic example of a persistent and malevolent chemical is DDT. Banned twenty years ago, it and its breakdown product, DDE, continue to show up in everything from fish to man, albeit in lower and lower amounts as time passes. Its rate of disappearance has been much slower than anticipated. Similarly, catastrophes like the Kepone spills into the James River in Virginia and PCBs in the bottom sediments of many rivers have persisted long after the initial contamination stopped.

We continue to add harmful compounds to our waters, particularly aromatic and chlorinated hydrocarbons. Pentachlorophenol, a common wood preservative, can contain a wide assortment of furans and dioxins as contaminents. Contaminated penta has been blamed for the recent collapse of a great blue heron population on Vancouver Island. The heron, an apex predator in the marine food chain, accumulates dioxin from fish. The dioxin causes reproductive failure in the herons.

Is the harm we have done irreversible? Yes and no. That we will continue to modify the world around us and seldom restore much of anything to its natural state suggests we have permanently lost most of our former wetlands. As for the poisoning of these waters, all it takes to clean them up is technology, wisdom, and the will to do it.

Can wetlands be reclaimed? Low-production marshland can be dredged deep enough to support the growth of high-productivity salt marsh cordgrass, but it is an expensive answer. However, developers have been willing to do it in exchange for building rights in sensitive areas. One such "mitigation" project is under way in New Jersey's Hackensack meadowlands. Environmental critics contend it is a poor trade-off in an already overburdened region.

Can shoreline development be contained? The prognosis

isn't good. Local control tends to fall into the hands of local businessmen, and new development means new business. Local failures to provide restrictions on coastal land use prompted the 1972 Coastal Zone Management Act (CZMA), which put the protection of estuaries, wetlands, dunes, beaches, and barrier islands squarely into the hands of the states. The carrot was federal funding. To qualify, states must carry out managing coastal development programs that meet federal criteria.

A few states have passed tough laws to help stop the decline of their waters. Maryland now has a series of restrictions on development, farming, logging, and mining within 1000 feet of Chesapeake bay tidewater and wetlands. Other states have run into heavy opposition from builders and landowners who claim such restrictions reduce the value of their property and are an encroachment on their legal rights.

Environmentalists have had high hopes for the long-term effectiveness of CZMA. This law places the care and responsibility for affected areas close to those who will benefit most, yet not so close that greed and shortsightedness will rule the roost. However, in most states, the jury is still out.

How we have allowed ourselves to drift into these dilemmas, the kinds of problems we now face, the extent of the damage, what the government can and cannot do, what concerned citizens are doing, what else needs to be done, and what *you* can do to help are all subjects addressed in subsequent chapters.

TWO

پایین رود

THE DOWNSTREAM
DELUGE جل

Every major waterway in the United States is part of the nation's sewage system. Not a city or a town with access to a stream or river has failed to make use of it as a conduit for its storm runoff or its municipal and industrial waste.

And all rivers eventually find their way to the sea.

What wasteloads reach the river's mouth depends on the terrain and urban centers through which they have coursed and the chemical and biological opportunities they have had to biodegrade those wastes while in transit.

In 1840, 89 percent of the U.S. population lived on farms. By 1880, 28 percent lived in cities of over 8,000 residents. Before the Civil War, both farm and town dwellers got their drinking water locally from wells, ponds, cisterns, and pumps. Waste washwater went into a dry well or a cesspool or was simply tossed out on the ground. On farms, human waste went into leachable cesspools. In towns it went into privy vaults, stone- or brick-lined containment vessels that had to be periodically emptied. The effluent, hauled away either by private contractors or by the city, at first found its way to local farms for fertilizer but, as cities grew and farms receded, it was merely dumped into local rivers.

H. L. Mencken, reminiscing about Baltimore in the 1880s, recalls their "domestic facility . . . was pumped out and fumigated every Spring by a gang of colored men who arrived on a wagon that was called an O.E.A.—i.e., odorless excavating apparatus." This process was accompanied by the burning of buckets of tar and rosin to swamp out the stench and was "thought to be an effective preventitive of cholera, smallpox and tuberculosis."

Baltimore's slops wound up in the Back Basin, a cul-de-sac whose waters circulated sluggishly into Chesapeake Bay. "As a result it began to acquire a powerful aroma every Spring and by August smelled like a billion polecats."

As city populations grew, this system of collecting waste simply proved inadequate. Overflowing and leaking sumps saturated the surrounding ground, flooded cellars, and threatened wells.

Back then, disease was thought to spread by unwholesome vapors emanating from filth. The connection between sewage pollution and illness was made well before Pasteur's germ theory of disease explained how this was possible.

During the nineteenth century, most major cities and towns constructed drainage systems for street runoff and developed public water supplies. Piped-in water was needed for fire fighting and to flush streets (recall that those were the days of horse-drawn commerce) as well as for domestic demand. None, however, concurrently developed the systems to carry away all the excess household waste.

The change that really overloaded the already antiquated horse-drawn haulage system was the widespread demand for the flush toilet, made possible by the availability of running tap water. Before running water was available, average daily water use ran about 3 to 5 gallons. By 1880, per capita use in Boston and Chicago exceeded 100 gallons per day.

Clean, odorless, and indoors, the attractiveness of the new water closet was irresistible, and by 1880 one-fourth of all urban houselolds had installed one. Initially they emptied into preexisting containment vaults, exacerbating what had already become a menace to health and an offense to the nose.

In 1850, the city of London had installed a combined sewage and storm runoff collection system that was to become the model for larger U.S. cities. The arguments over how best to construct sewer systems were based on a series of misconceptions. Disease was believed to be carried by "sewer gas." Since sewer gas formed more readily in combined storm and sewage systems, a number of smaller cities and towns opted for separate lines. Large cities, faced with enormous capital costs for construction of dual pipelines, cited the London model and combined their effluents with street runoff.

By 1909, more than twenty-five thousand miles of pipeline had been laid in cities of over 30,000 population. Slightly over 20 percent were separate sanitary sewers.

Initially, raw sewage was dumped into the nearest convenient body of running water. The effects of this on the drinking water supply of towns downstream was largely ignored because it was widely held that "running water purifies itself."

Between 1890 and 1910 this assumption was challenged, in part because of the rapid rise in waterborne diseases, particularly cholera and typhoid fever. Typhoid struck hard in Atlanta and Pittsburgh, which had both constructed sewer systems but had failed to protect their drinking water from sewage contamination.

Early sewage treatment was undertaken more to avert the nuisance it created than to eliminate its health risks. In 1900 the two most prevalent treatment methods, sewage farming and intermittent filtering, required large tracts of land, a luxury that cities were rapidly losing. Neither method could handle the volumes produced by combined systems.

About then, purifying drinking water by filtering it through sand began to reduce the incidence of waterborne diseases. The issue for municipalities then became whether to treat waste, treat drinking water, or do both.

The fear of epidemics, their continual recurrence, and the toll they took brought enormous public pressure to bear on state and local authorities to improve drinking water purity and to compel cities and towns to stop discharging untreated waste.

Between the turn of the century and the first World War, there raged between upstream and downstream municipalities hundreds of disputes that occasionally erupted into bitter controversies between adjacent states.

The upshot of these arguments was that emphasis shifted from treating sewage to filtering and chlorinating drinking water.

On the eve of World War II less than 60 percent of sewered waste received any treatment at all. During and after the war, industry and population expanded rapidly. The infrastructure to support this growth was simply added on to what already existed, whether or not it was appropriate to the ever-increasing sprawl.

Primary sewage treatment became mandatory and the septic systems of less populous areas were abandoned. Today, 30 percent of the nation's publicly owned treatment works (POTWs) still go no farther than primary treatment; the bulk of the remainder render secondary treatment. Both terms are explained below.

WASTE TREATMENT

Today, essentially two kinds of collection systems exist. One pipes sanitary and industrial waste through one system and handles storm runoff through a separate network. (The waste goes for treatment; the street runoff goes directly into the nearest river or stream without further ado.) The other combines the two in a single system, where both waste and runoff are treated. When it rains, the volume of combined sewage exceeds the capacity of the treatment plant and raw sewage is directly diverted to the receiving stream.

In primary treatment, the POTW may do little more than screen out debris, separate stones and grit, settle out some suspended solids (raw sludge), and chlorinate the remaining water to kill bacteria.

In secondary treatment, the waste is further decomposed by bacterial attack. The two main methods in use today are the trickling filter, in which the waste is sprayed over a deep bed of stones that have a film of bacteria growing on them, or by the activated sludge method, where the waste is mixed with activated (bacteria laden) sludge and recirculated through a tank that is violently agitated with air.

In regions where groundwater is scarce and land is available, tertiary treatment allows the effluent to be directly recycled back to the soil. The system involves, among other things, holding the waste for thirty days in large, shallow lagoons.

The main goal of the municipal treatment plant is to reduce the biochemical oxygen demand (BOD),* the total suspended solids (TSS), oily substances, and bacteria content of the discharge. Primary treatment removes 60 percent of the solids and 30 percent of the BOD; secondary treatment removes 85 percent of both BOD and TSS. Tertiary treatment reduces BOD and TSS by over 95 percent and significantly reduces nitrogen. Neither primary nor secondary treatment removes significant quantities of nitrogen or phosphorus.

*BOD is a measure of the amount of dissolved oxygen required by bacteria and other microorganisms to oxidize organic wastes in water within a specific period (five days is the common assay time).

The most common method of destroying bacteria in effluent is chlorination. Chlorine is cheap, easily handled, and a powerful disinfectant.

Sewage chlorination has its drawbacks. It destroys fecal coliform bacteria reasonably well but is not so effective against a host of other bacteria and viruses. Chlorine reacts with a number of substances in waste and can form chlorinated hydrocarbons harmful to organisms living near the point of discharge.

What the sewage plant manages to separate out from the effluent is called sludge. Heavy metals, organics, oil, grease, bacteria, viruses, and protozoans all accumulate in sludge. If the sludge is to be put on land it must be "conditioned" (dewatered, digested, and disinfected). If it is to be used for fertilizer, the EPA requires that it must be subjected to high enough temperatures to destroy common pathogens.

The efficiency of public waste-treatment plants can be undermined by events over which the plants have little control. Sudden deluges of industrial waste containing toxic concentrations of metals or organics can kill or inhibit the growth of the bacteria that make the plant operational. These "upsets" cannot be remedied; the waste is simply pumped out untreated. Higher-than-normal waste loadings can also overload a plant. Few plants have excess capacity to deal with sudden surges.

Many plants intermittently or continuously violate EPA compliance standards. Tight budgets, municipal indifference, and bureaucratic mismanagement have too often resulted in broken-down and outmoded equipment as well as poor plant maintenance and operating procedures.

Of the 15,500 or so municipal sewage treatment plants now in existence only 578 discharge directly into marine waters. However, this small number accounts for 25 percent of the nation's total annual output of 9.5 trillion gallons of waste.

Of those 578 plants, 509 discharge directly into estuaries to the tune of 2 trillion gallons per year. The remainder discharge 0.3 trillion gallons per year into coastal waters. About one-seventh (0.33 trillion gallons per year) of the wastewater handled by treatment plants that discharge into marine waters originates from industrial sources.

Industrial wastewater can legally be discharged either directly into a receiving stream (whatever waterway the outfall pipe dumps into) or indirectly, through a public treatment plant.

Excluding power plant cooling water, 1317 industrial pipelines discharge just over 4 trillion gallons per year of effluent directly into estuaries; only 15 facilities discharge into coastal waters.

Combined municipal and industrial discharge into coastal rivers adds another 4 trillion gallons annually. About 60,000 direct industrial dischargers, as well as the 15,000 public waste treatment plants, are regulated and must comply with federal and state standards. Over 130,000 industrial and commercial businesses are indirect dischargers that send their waste into public treatment plants. Indirect dischargers include laundries, printers, timber processors, plastics molders, paint and ink manufacturers, and textile mills. Only about 15,000 of these are covered by categorical standards.

Treatment plants with daily flows exceeding 5 million gallons and smaller ones that receive significant quantities of industrial waste must, by law, develop and implement individual pretreatment programs, but only 1500 of the 15,000 publically operated treatment plants have done so. However, those 1500 receive 82 percent of the total industrial wastewater entering all such plants and receive 90 percent of the waste that must meet categorical pretreatment standards.

Direct dischargers span a wide spectrum, and include everything from coal and ore mining, iron and steel production, and petroleum refining to chemicals and pharmaceuticals manufacture. Each direct discharger falls within an EPA category for which effluent standards have been set or will be.

Total industrial wastewater output is over 6.4 trillion gallons a year. About three-fourths of that exits directly into receiving waters, the rest flows through treatment plants.

MAJOR POLLUTANTS

The Clean Water act of 1972 classifies pollutants as conventional, nonconventional, and toxic. The conventional municipal pollutants are TSS, BOD, oil and grease, pH and fecal coliform bacteria. Nonconventionals are a catchall category that includes the major sewage nutrients, nitrogen, and phosphorus, as well as other substances. Toxic pollutants include environmentally harmful metals and organics. EPA currently lists over 126 such toxic "priority" substances which it wants substantially reduced through pretreatment.

Conservatively, industrial waste streams discharged 600 million pounds of these priority metals and 230 million pounds of organic chemicals annually before the pretreatment program began. These numbers are being reduced, but the extent of overall reduction is not accurately known. Some industries, like metal finishing, electroplating, and pulp and paper, are seriously lagging, while others, such as organic chemicals, plastics, and synthetic fibers, have achieved a 99 percent reduction in pollution loadings.

Toxic Organic Chemicals

Over 65,000 organic chemicals are commercially available. Of these, 10,000 are used regularly in one or more industrial processes.

The toxicity of synthetic organic chemicals varies widely. Some decompose rapidly when exposed to air, sunlight, and water. Others are persistent, resisting decomposition by natural processes. Of those, especially compounds with a propensity to dissolve slightly in water and more readily in fats and oils, a few can bioaccumulate in living tissue. An animal that ingests such a compound may or may not be harmed by it. When the host animal is preyed upon, the compound can be passed on to its predator. DDT and a number of other chlorinated hydrocarbons can follow such a pathway.

Of current concern are the polychlorinated biphenyls (PCBs), polyaromatic hydrocarbons (particularly naphthalenes and benzopyrene), chlorophenols and chlorobenzenes, and the chlorinated pesticides such as Kepone and Mirex.

DDT and PCBs have become widely distributed in sea life. How DDT got around is now well known, but the distribution pathways and effects of PCBs have proved to be more elusive. PCBs are relatively inert mixtures of chlorinated biphenyls that are poorly soluble in water. Some 210 different compounds are theoretically possible. The average industrial PCB contains 40 to 60 percent chlorine. PCBs with high chlorine content are more persistent, but less toxic, than those of lower chlorine content. Mammals and birds are less affected by PCBs than are fishes and invertebrates. Toxicity tests on a variety of animals have been confounded by considerable variation among samples of product from various manufacturers. Some sources contain traces of dibenzofurans, extremely toxic and carcinogenic compounds closely related to the dioxins, which apparently are formed during synthesis. These may be the cause of much of the apparent toxicity of PCBs.

The only case of human illness directly attributed to PCB have resulted from contaminated cooking oils. In southern Japan, in 1968, more than a thousand persons who had eaten

19

contaminated rice oil developed dark skins, eye discharges, severe acne, and other symptoms that came to be known as Yusho—oil disease.

Concern has also centered on compounds that cause tumors and lesions in laboratory animals and can find their way through the aquatic food chain into humans. Again, the chlorinated hydrocarbons and polyaromatic hydrocarbons are major suspects.

Polyaromatic hydrocarbons, including the classic carcinogens benzopyrene and benzanthracene, occur in crude petroleum. Both crude and refined fractions of petroleums from a number of sources, however, are relatively nonmutagenic in the commonly employed Ames test. Used crankcase oil is a different story. It is mutagenic and is probably a major ecological threat to aquatic life.

A host of other aromatic hydrocarbons are found in sewage. A metabolite from the breakdown of nonionic surfactants, called 4-nonylphenol, has been found in concentrations that are extraordinarily high compared to other contaminants in digested sewage sludge. This compound is both toxic and mutagenic.

Just how pervasive some organic chemicals are is exemplified by the widespread distribution of phthalate ester plasticizers in the marine environment. These compounds are put into plastics to give them flexibility and are rarely used elsewhere. They are poorly soluble in water, absorb tightly to sediments, and are not biologically degradable. They have been found in the waters of the North Atlantic and in the waters, sediments, fish, and invertebrates of the Gulf of Mexico and, in particular, sediments in the Mississippi delta.

Thousands of organic chemicals enter our municipal and industrial sewage systems every day. In one survey, EPA identified 385 compounds (with hundreds of others unidentified for a variety of technical reasons) in wastestreams. Six of those, toxic and not treatable—dibenzofuran, two trichlo-rophenols, carbazole, trichlorobenzene, and a form of dioxane—were present in significant amounts.

Although it has been hard to correlate specific organics in waste with human afflictions, circumstantial evidence makes our wastestreams suspect. The Mississippi River collects an abundant wasteload on its way to the sea. Parts of Louisiana use it for drinking water. Those that do, experience statistically significant higher rates of cancer—total cancer, cancer of the urinary organs, and cancer of the gastrointestinal tract—than those who do not.

Toxic Metals

There is no clear-cut, unequivocal way to define a poisonous natural element or a toxic metal. The physiological action of an element depends on its form and concentration as well as its latent potential for mischief. Barium, in ionic form, is a poison—yet barium sulfate is so insoluble that it can be swallowed in large quantities with impunity. Arsenic, chromium, cobalt, copper, molybdenum, nickel, selenium, and zinc are considered toxic metals, but all are also essential trace elements for humans and, presumably, other life.

For a toxic metal to be harmful, it must be accessible to animals or plants in an assimilable form at significant levels. Elemental mercury and its less soluble inorganic salts must first be converted by microorganisms into methylmercury before it can be absorbed by molluscs or fish. A certain background amount of mercury constantly recirculates through the biological cycle in this manner. If much higher quantities of mercury enter the cycle at a specific location, accumulations in marine life in that area can cause serious consequences to humans. In Japan from 1953 to 1958 over one hundred people died and over seven hundred others suffered severe, permanent nerve damage after eating shellfish contaminated with industrial waste containing methylmercury that had been dumped into Minamata Bay.

The Japanese have also endured other metal poisoning episodes. A mining operation that discharged cadmium-bearing waste into the city of Toyama's Jintsu River, which supplied water for rice paddies and soybean cultivation, produced a chronic affliction called "Itai, Itai." More than two hundred people developed symptoms that mimicked rheumatism and neuritis, followed by bone decalcification and kidney failure. Over half its victims died.

Cadmium is widely used in electroplating and commonly contaminates sewage waste, sludge, and marine sediments near urban areas. The Hudson River estuary is an example. Shellfish harvested near one industrial outfall formerly discharging cadmium waste contained enough of that metal to induce acute poisoning in anyone unlucky enough to eat them.

The four metals that primarily concern EPA presently are arsenic, cadmium, lead, and mercury. Both toxic and in relatively high concentrations in waste, these elements have the potential to be biologically recycled into humans. Chromium, copper, and selenium, found in lower concentrations, have a similar capacity but are less poisonous.

Primarily found in sludges as insoluble compounds, these elements can only enter the food chain and cause harm if they are converted in either water- or oil-soluble compounds. Like mercury, arsenic and lead can be methylated by certain bacteria. These methyl derivatives can then be taken up by other living things. Cadmium cannot be made soluble by methylation but can be solubilized by oxidation or chelation. Chelators, such as EDTA or NTA, commonly used as water softeners, can react with a host of heavy metals, some benign and some not, and convert them into water-soluble compounds.

Neither arsenic nor lead has been shown to accumulate in seafood. Although metals can bioaccumulate in an organism, most do not biomagnify, that is, they do not increase in concentration progressively in an animal that preys or feeds upon another organism that initially ingests the metal. However, mercury, selenium, and zinc are exceptions to this rule. High levels of selenium have been found in the tissues of two species of marine birds—Texas laughing gulls in Galveston Bay and California scoters in San Francisco Bay.

The sources of toxic metals in sewage waste in urban areas are not wholly industrial. The city of New York found that its residential contribution accounted for 47 percent of the copper, 49 percent of the cadmium, 42 percent of the zinc, and 28 and 25 percent of chromium and nickel. Electroplaters contributed the bulk of the chromium and nickel, 43 and 62 percent respectively. Street runoff contributed about 10 percent of the metals mentioned except for zinc (31 percent).

Human Pathogens

Pathogens, microbial organisms that can cause disease, enter the marine environment via raw and poorly treated sewage discharges, in sludge, and from leachates and runoff.

In the United States, bacteria and viruses are the main health concerns. Protozoans that cause dysentery, parasitic flatworms that cause trichinosis and schistosomiasis, nematodes and cestodes that cause roundworm, threadworm, and tapeworm infestations have proven less worrisome.

The principal bacteria commonly found in sewage are as follows: *Escherichia coli*, the fecal coliform found in all of us, and which can cause enteritis in children; *Salmonella*, responsible for typhoid and paratyphoid fevers, food poisoning, and gastroenteritis; *Shigella*, bacterial dysentery; *Clostrid-*

ium, food poisoning; *Staphylococcus aureus,* wound infections and food poisoning; *Mycobacterium,* tuberculosis; and *Leptospira,* Weil's disease and jaundice.

The viruses include poliovirus, infectious hepatitis virus, adenoviruses that cause eye and pharynx infections, coxsackieviruses A and B, ECHOviruses and others that cause enteritis, fever, rashes, and central nervous system disorders.

Bacteria and viruses survive in the marine environment for many months. An outbreak of cholera along the Gulf coast of Texas appears to have been caused by bacteria that have persisted for at least five years in coastal waters.

Human pathogens seem to survive better in estuaries and coastal waters than in the open ocean. They collect on particulates and can be found both on suspended matter and in bottom sediments.

For bacteria and viruses to infect humans they must first reach them, and do so in appropriate numbers. The greatest threat is through infected shellfish. Shellfish have been implicated in outbreaks of typhoid and parathyphoid fevers, hepatitis, and, most commonly, viral gastroenteritis. In 1983, New York State identified shellfish consumption as the cause of 103 incidents of gastroenteritis involving over one thousand people. Shellfish have also been implicated in cholera outbreaks along the coast of southwestern Louisiana.

Viral gastroenteritis and hepatitis have also been traced to swimming in polluted water. The number of incidents has been steadily rising in recent decades. Epidemiological studies over the last fifteen years have yielded mixed results. Some claim the risk of swimming in grossly polluted water is no greater than in a well-chlorinated pool; both situations result in higher occurences of ear, eye, nose, throat, and stomach ailments than for nonswimmers.

THREE

THE BELEAGUERED ESTUARY

On long stretches of coastline fresh water and salt merge gradually in an irregular and blurred edging. This transitional zone and its adjacent waters make up the estuary, the richest of all marine ecosystems.

Its output surpasses our best farmland. An acre of grain fed to livestock will yield just over 20 pounds of dry weight protein in a year, but fish and prawn ponds in third-world countries regularly produce twice that amount. Simple oyster culture yields 100 to 150 bushels an acre. In the United States, two-thirds of the cash value of commercial catches depends on fishes which occupy estuaries for some part of their life cycle. For recreational fishermen these waters supply the bulk of their catch.

These brackish waters are home to our most valued shell-fish: oysters, bay scallops, soft-shelled and hard-shell clams. They are a way station for the fishes whose lives are divided between fresh water and salt, like the salmon and shad. They are regularly visited by callers from the open sea: atlantic mackerel, menhaden, bluefish, and squid. Some migrants, such as croaker and spot, spend their summers in the estuaries and winter offshore; others, winter flounder for example, do just the reverse. Estuaries are the nursing grounds for a cornucopia of fish and invertebrates. Spawned there, some, such as striped bass, weakfish, or drum, grow up and stay close to shore for the rest of their lives. Others, spawned offshore, make their way to the shallows to feed and grow to maturity, among them the flounders, scup, and prawns.

An estuary has no clear-cut definition. It is a confined body of water protected from the open sea and sheltered from heavy wave action, shallow enough in places for grasses and kelps to prosper, and flushed by freshwater runoff and tidal action. It is a collection of waters of mixed salinities over a variety of bottoms: marshes, wetlands, bays, reefs, embayments, flats, deltas, and bars.

CARRYING CAPACITY

The ecosystem of any estuary provides essential elements that allow life within to proliferate. The foundation of this "carrying capacity," as it is called, rests upon its capacity to produce plant life, known to ecologists as its "primary productivity." In turn, primary productivity is regulated by the extent to which sunlight, carbon dioxide, and nutrients combine to produce attached and floating plant life.

Plants in the form of phytoplankton, are fed upon by zooplankton and larger filter feeders. Foragers prey upon these microscopic animals and are themselves fair game for predators like large fish, birds, and mammals. At the end of the chain are the decomposers—bacteria and plankters—that convert detritus back into elementary compounds.

Grasses and kelps suffer little direct grazing. Their major role lies in their natural death and renewal. This forms the basic storehouse from which nutrients are drawn and deposited in a perpetuating cycle. Because the cycle is slow, these plants cushion against wide swings in the availability of nutrients. They create a buffer against times of want.

Beyond their biological value, their physical presence creates vast sediment traps that hold, bind, and then use the drift of mineral particles washed off the land. These particles maintain and create a substrate for root systems and supply them with essential elements. This unique capacity to store and slowly release organic and inorganic material, as well as to capture and convert the sun's energy into a usable form, is the cornerstone of estuarine productivity.

The carrying capacity of the estuarine ecosystem depends on the physical and chemical condition of its waters. Water circulation—offshore currents, tidal action, freshwater inflow, and wind—controls the arrival of oxygenated water, the flushing of wastes, the transport of plankton and larval forms, and the movement of nutrients.

Clear water enhances carrying capacity. Sunlight is the source of the ecosystem's energy. To use it, the water must be clear enough to allow light to penetrate to bottoms where grasses grow. Cloudiness in the water, from either suspended silt or small organism overproduction, will cut estuarine production, and continual turbidity will choke off the growth of bottom grasses, notably eelgrass in temperate climes and turtle grass in the south.

The chemical makeup of these waters is critical; the presence of too few or too many nutrients spells trouble. Nutrients and trace elements brought in by rain or runoff or from offshore are vital to plant production. Nitrogen and phosphorous are key to plant growth. Plants require more nitrogen than phosphorus, and, in coastal waters, nitrogen is usually the factor that limits their abundance.

Estuaries generally contain wide ranges in salinity that

shift seasonally with differences in upstream rainfall. Varied salinity within the estuary increases its habitats. Some species can tolerate a wide salinity range; others cannot. Salt marsh grass prefers salinities of over fifteen parts per thousand, whereas the blue crab can get along equally well in water either nearly fresh or very salty. Salinity differences often keep predators and prey apart. The atlantic oyster can tolerate a wide salinity range, but its major predator, the oyster drill, is confined to waters over fifteen parts per thousand.

Certain families of annelids (segmented worms), molluscs, and crustaceans have adapted so well to low and medium salinities that they are represented in estuaries worldwide. Nereids (of which the clam worm is an example), bivalves (especially clams, oysters, scallops, and tellins), snails, and crustaceans (amphipods, isopods, shrimp, and crab) can be found in brackish water nearly everywhere.

Salinity partitions seaweeds, aquatic weeds, wetland plants, invertebrates, and fish within the confines of the estuary. In temperate waters, sturgeons, eels, some herrings, shad, mummichog and silversides, white perch, and striped bass are equally at home in sweet water or salt. Gars, pike, suckers, crappie, and smallmouth and largemouth bass rarely venture downstream into salinities of more than a few parts per thousand while shark, rays, sea bass, jacks, grunts, porgies, kingfish, blackfish, sea robin, and filefish penetrate no farther than the 18 to 30 ppt (parts-per-thousand) contours.

Between the two extremes (5 to 20 ppt) you are apt to find minnows, carp, shiners, toadfish, halfbeak, killifishes, sticklebacks, sunfishes, bluefish, sea trout, weakfish, spot, croaker, mullets, blennies, gobies, mackerel, flounder, soles, and puffer.

The water's capacity to dissolve carbon dioxide and oxygen is critical to all life within it. Plants use carbon dioxide and release oxygen, whereas animals do the reverse. Because bacterial decomposition and temperature also consume or limit oxygen, the oxygen levels in the estuary change seasonally, falling during the summer. Should those levels fall below minimum needs, disastrous consequences result. At 6 parts of oxygen per million, all is well; below 4 ppm certain animal life begins to suffer; at 2 ppm, fish are in serious trouble; below 2 ppm, for any sustained period, widespread suffocation takes place. At zero oxygen, new processes take over. Aerobic decomposition turns anaerobic (without oxygen), with the subsequent production of hydrogen sulfide. Some animals can tolerate this for a short period of time, simply by stopping breathing, but this substance is so poisonous that few creatures can put up with it for long.

The coastal watershed surrounding the estuary strongly influences its well-being. High marshes, woodlands, and freshwater wetlands trap and relinquish runoff gradually, absorb pollutants, and release detritus useful to the marsh and lagoon dwellers. This shoreland sets up a buffer zone, moderating the volume and quality of the water that passes through it.

The balance of these factors—salinity, freshwater inflow and circulation, nutrient loading, dissolved oxygen, water clarity, storage capacity, and the watershed screen—establish the carrying capacity of the system.

STRESSES ON THE ESTUARY

Life in the estuary is subject to considerable natural stress. The environment is unpredictable. Severe fluctuations in salinity and oxygen make life tenuous for all but the most resilient and hardy. As a result, the estuary does not have the species diversity of the ocean nor that of fresh water. Natural events like the 1972 tropical storm, Agnes, that hit the Chesapeake Bay estuary and caused the most severe floods in two hundred years, upset the whole ecosystem from the Sassafras River in the north to the mouth of the bay.

In a number of U.S. estuaries wide swings in population densities of bottom dwellers are commonplace. In others, some species persist in the same location year after year, while others fluctuate seasonally or go through brief, sharp increases in numbers followed by longer term quiescence. These temporal swings may be the result of a hard winter, heavy predation, the invasion of a competing species, or a subtle change in the surroundings.

The activities of man have drastically altered our major estuaries and have added additional stresses to boot. These changes have invariably been for the worse.

The rapid encroachment of cities and towns right up to these waters, coupled with the proximity of farmlands now heavily dosed with chemical fertilizers and pesticides, has removed the isolating effect that the coastal uplands once gave. Now unfiltered runoff brings with it not only an overload of nutrients and toxic waste, but also turbidity, which chokes off bottom life and blocks the sun from the bottom grasses, reducing still further the storage capacity of the system.

The loss of wetlands and salt marshes has undermined the great wealth of the estuary. Half of the nation's wetlands are gone, filled in by everything from dwellings to dumps. Half of what remains has been impaired by the degradation of their surroundings.

The yields of the estuary to humans has plummeted over the past half century. New York, New Jersey, and Connecticut have experienced an 80 percent decline in their estuarine-dependent commercial fisheries. Chesapeake Bay output has fallen dramatically in past years and continues to decline.

Fresh Water Flow

The flow of fresh water to many estuaries has been diminished, by either damming or diversion, permanently changing the salinity in the estuaries downstream. Examples are many. The Florida Everglades, an enormous cedar, red mangrove, and saw grass wetland, is the primary nursery grounds for pink shrimp, gray snapper, red drum, snook, and spotted sea trout. Its productivity hinges upon the flow of fresh water from the north. A system of levees and canals now controls its flow and that flow depends entirely on the decisions of the U.S. Army Corps of Engineers and the South Florida Water Management District, which, in the past, have laid more emphasis on the agricultural interests to the north than on the wildlife to the south.

Channel dredging, diking, and filling have altered estuarine circulation patterns, upsetting the dispersion of drifting life and the distribution of nutrients. An especially egregious misuse of wetlands has been lagoon development for residences. These create dead-end, bulkheaded canals, deeper than the estuary waters around them. Like open cesspools, these poorly circulating enclosures quickly foul from seepage.

Warm-Water Effluent

Dissolved oxygen in estuaries has been reduced by the vast loads of organic waste, both dissolved and suspended, swept in by sewage and soil erosion. Some estuaries are also burdened with warm water discharged from power plants, which not only reduces the water's capacity to carry oxygen but upsets the reproductive cycle and migration patterns of many of its inhabitants. Warm-water effluents from utilities have mixed effects on life in their immediate vicinity. They can prolong the season and induce rapid growth in some organisms but, on the whole, their influence is destructive.

Power plants' requirements for cool water are enormous—1000 to 5000 cubic feet per second, the flow of a medium-sized river. The intake passes through a screen to prevent entrapping large objects, into the pumps, and on to the heat exchanger tubing, and is then discharged. The temperature of the water may rise 10–34° F during the brief trip.

Small fish and larger plankton are trapped against the screens. One operating unit was killing five million fish a week before the intakes were modified to prevent it. Most of the smaller plankton passing through the screens are killed by the mechanical pressure of the pumps. The water flowing into the discharge canal has a significantly higher BOD* than the intake water. The Oyster Creek nuclear generating station, which discharges into Barnegat Bay, New Jersey, grinds up enough living protein each day to require 17,000 pounds of dissolved oxygen to decompose it, a BOD loading equivalent of a 25 million gallon a day primary sewage-treatment plant.

The discharge canal becomes a scavenger habitat. During the summer months, phytoplankton in the thermal plume slow their respiration and rate of photosynthesis. Adult finfish avoid the plume but, as winter closes in, a number of migratory species linger in the warm water. All generating plants require periodic shutdowns, some scheduled and some not. When the plume stops in cold weather, the fish in it die of thermal shock.

Free Chlorine

Some plants chlorinate cooling water to prevent fouling in the condenser tubes. About 0.25 to 0.75 parts per million free chlorine is needed. At 0.1 ppm, phytoplankton respiration

---

*Biochemical oxygen demand, the amount of dissolved oxygen required by microorganisms to oxidize the organic waste in that water.

and photosynthesis are inhibited by 80 percent and protozoans die. Within six minutes in such water, fish become disoriented and swim haphazardly.

Excess chlorine in effluent discharge is mainly a matter of poor dosage control. Overchlorinated sewage, cooling water, and other chlorinated effluent, such as pulp mill waste, can be detoxified with one of several easily handled reducing agents: sulfur dioxide, sodium sulfite, or thiosulfate.

Metals

Heavy metals get into the food chain either by the direct ingestion of sewage or sediments or by absorption from sewage-laden waters. Heavily contaminated bottom sediments, commonly found in harbor mouths, can contain very high amounts of metals. In Newark Bay, New Jersey, levels for zinc are 65 times normal, and 128 times for lead, 180 times for cadmium, and 155 times for mercury.

Estuarine molluscs accumulate metals; the American eastern oyster and the Pacific oyster do so efficiently enough

Range of Metals Values in Oysters
(ppm wet weight)

	EAST COAST, U.S.	WEST COAST, U.S.
ZINC	180–4120	86–344
COPPER	7–517	8–37
IRON	31–238	15–91
LEAD	0.1–2.3	0.1–4.5
CHROMIUM	0.04–3.4	0.1–0.3
CADMIUM	0.1–7.8	0.2–2.1

Source: Pringle et al, 1968.

to be useful as indicators of metal pollution. (Both can accumulate enough zinc and copper to give their flesh a blue-green color.) Oysters collected from Maine to North Carolina and from Washington on the west coast show the extent of metals pickup.

The variation in iron is partly due to natural sources, but the higher values of the other metals are from sewage or runoff.

If oysters are moved to metal-free water, they do not readily rid themselves of copper or cadmium. Even low levels of cadmium in the water interfere with larval development. Mercury, copper, and zinc are also toxic to oyster larvae in far lower concentrations than those which affect adults. The same can be said for a host of crustacean larvae.

Seafood contributes about 10 percent of human exposure to cadmium and is the major source of exposure to mercury. Mercury contamination occurs from "hot spots," more often than not former sites of chemical plants manufacturing chlorine by the "mercury cell process" or other processes. The Waynesboro, Virginia, plant that DuPont operated for the manufacture of acetate fibers between 1929 and 1950 has continued to pollute the south fork of the Shenandoah River for thirty-nine years, with no end in sight. An Olin chlorine plant located in Saltville, Virginia, on the Holston River, and closed since 1972, has in its soil more than 220,000 pounds of mercury that is leaching into the river at a rate of a little over a pound a week.

Acid Rain

Until recently, acid rain had been thought to have little impact on marine waters, but a recent study by the Environmental Defense Fund calls acid rain a major contributor of nitrogen in eastern estuaries, particularly Chesapeake Bay. Oxides of nitrogen are emitted by coal and oil burning utilities and by automobiles. Nitrogen oxides quickly oxidize further and combine with water vapor to form nitric acid, which falls with the rain.

Oil

Perhaps the most ubiquitous of all contaminants in the harbors and estuaries of the United States is oil.

Oil has been damaging these waters since the Civil War. In 1887, George Goode of the Bureau of Fisheries conducted a census of eastern fish stocks and observed that spills from Newark Bay refineries were disrupting and reducing the distribution and abundance of local shellfish and finfish. The reduced catches were unmarketable; they tasted of coal oil (kerosene).

In 1912, the New York Zoological Society, which then operated the New York Aquarium at Battery Park (now housed at Coney Island), reported it could no longer use local harbor water for its tanks because oil contamination was killing specimens. In 1920, a local researcher noted that many of the molluscs once common to the Hudson River and Staten Island had disappeared; again the culprit was oil.

Marine ecologists generally agree that chronic low levels of oil are more damaging to marine life than single catastrophic spills. Nevertheless, a single small spill in a marshland can do immediate and long-term harm.

In September 1969, the barge "Florida" spilled six hundred tons of diesel oil into Buzzard's Bay, off West Falmouth, Massachusetts, practically in the backyard of the Woods Hole Oceanographic Institute. The oil took a heavy toll of all the life it came in contact with. Within a week bottom animals declined from two hundred thousand organisms per square meter to two per square meter. Fish and lobsters died within hours. Two days later, in Wild Harbor River and West Falmouth Harbor, marsh grass died as did clams, oysters, scallops, and small invertebrates.

The oil penetrated into the sediments, as deep as five feet in places. A month later, the affected intertidal area was a biological desert. Dr. Howard Sanders, a Woods Hole ecologist who followed the effects of the spill, remarked "You could go down to the marsh and there wouldn't even be gnats. No mosquitos, no greenflies, no nothing . . . there was nothing alive at all."

Three years after the spill, clams in oiled areas were not breeding and some marshes remained closed to all shellfishing. After seven years, the effects could still be discerned on the fiddler crab: decimated by the spill at first, their recovery was inhibited by the residual oil in the sediments. Not only was that oil directly toxic in concentrations of 1000 parts per million to adults and 200 ppm for juveniles, but at lower concentrations (100–200 ppm) it also caused behavioral changes and locomotor disorders. Affected crabs dug shallow, irregular burrows and were slow to respond to threats.

A spill in Maine, where the oil had worked down into sandy bottom sediments, kept killing soft-shelled clams three years after it happened. The clam larvae settled normally and grew but ran into trouble when they began to burrow and encountered the oil layer.

A 250,000-gallon spill of heavy (number 6) fuel oil in February 1976 fouled beaches and marshes on the eastern shore of Chesapeake Bay in Northhampton County, Virginia. As with the Falmouth spill, the shoreline was quickly cleaned up. Unlike the Falmouth spill, the reported damage was minimal, because the oil was much heavier than that involved in the Falmouth incident. Generally speaking, the heavier the oil, the less toxic it is. Mussels and oysters survived. The periwinkle, *Littorina irrorata*, that lives on the salt-marsh cord grass, *Spartina alterniflora*, was initially wiped out but recovered within eight months. The marsh grass actually increased in numbers and size during the following growing season.

BIOLOGICAL CHANGES

That life persists in our most polluted waters is a tribute to its adaptability. But even where life forms hold on, pollution brings biological changes that serve as warning signals to its presence. The most obvious symptom is mass mortality. Fish and shellfish kills have grown more frequent in bays and river mouths. Natural causes can trigger fish kills (red tide is an example), but their growing frequency in polluted areas points the finger toward man-made origins.

Some animals thrive in sewage. The asiatic clam, *Corbicula,* will proliferate right into outfall pipes and grow so rapidly it will clog them. In fresh water the sludge worm, *Tubifex tubifex,* dominates the life in sewer sediment. In salt water, polychaete worms assume that role—the very small *Capitella capitata* in California and *Heteromastus* and *Nereis* in temperate eastern waters.

A shift in bottom-dwelling species is another sign of deteriorating water. Brown algae and grasses diminish; blue-green algae and sea lettuce prosper. Nematodes, nemerteans, small polychaetes, oligochaetes, and other "worms" take over.

Abnormalities and disease in fish have increased over the past thirty years. Tumors, "fin rot," lesions, and the spread of virus diseases like lymphocystis, now endemic in estuaries along the Gulf of Mexico, are attributed to pollution.

Some afflictions of fish have been traced to chemical contamination. Scoliosis, a lateral curvature of the spine, c caused by dietary deficiencies or parasitism but al ganophosphate and carbamate poisoning, heavy metals, and Kepone. The contamination of the James River with Kepone has long since shut down the commercial fishery from Richmond to Chesapeake Bay.

Egg and larvae abnormalities in fish, fungus infections and exoskeletal erosion in crustaceans, and odd growths on sessile invertebrates have all been attributed to chemical contamination. Pesticides brought in by runoff are a major source. Chemical pesticides are meant to kill at very low concentrations. The more successful ones now in use, the organophosphates, are neurotoxins. Unfortunately these compounds are not specific to their target species alone. Very low levels kill or stun fishes, alter their blood chemistry, tissue metabolism, and ovary development, and disrupt their feeding and schooling behavior. These compounds can upset mollusc development, inhibit shell growth in oysters, destroy eggs and larvae of oysters and clams, kill crustaceans, and diminish phytoplankton populations.

Algal Blooms and Red Tides

Algal blooms and "red tides" have been increasing worldwide. Until recently, scientists ascribed these to natural perturbations, but their appearance in new locations and continual recurrence have raised suspicions that excess nitrogen is a prime trigger in setting them off.

An outbreak can discolor the water green, brown, yellow, or red depending on the species involved. Their numbers skyrocket until they become so dense they block off the light and consume all the nutrients. Then they begin to die off. The collapse of the bloom and their decay deplete the dissolved oxygen in the water, suffocating whatever is unlucky enough to be trapped nearby.

If the proliferating organism is a particular dinoflagellate, another risk arises. These small, one-celled creatures manufacture a deadly neurotoxin. Molluscs who feed on them can concentrate this toxin in their tissues without ill effect. However, if fish or man eat the afflicted mollusc, they can suffer paralytic shellfish poisoning, a potentially fatal illness. Authorities generally shut shellfish beds during these "red tide" incidents for fear of this possibility.

Bloom organisms can degrade the environment simply by crowding out all other forms of microscopic life." In the relatively clean waters of Peconic Bay at the easternmost tip of Long Island, three years (1985–87) of huge blooms of a minute brown alga named *Aureococcus anorexefferens* has wiped out the bay scallop industry. The scallop cannot capture and digest the algae (it is too small) and no other plankton are available for food. The bloom has been so thick it has blocked out sunlight to the bottom, killing off entire beds of marine grasses. The disappearance of these grasses in turn has reduced the volume of fish fry that attract so many migrant species of fish (bluefish, mackerel, flounder and fluke) to the bay in spring and fall.

A textbook case of the effects of pollution on algal blooms in shallow waters came in the growth of the Long Island duckling industry. Duck farmers along the rivulets feeding Moriches Bay placed their pens adjacent to (and in) tidal water. The droppings were swept into Great South Bay and Moriches Bay, two contiguous embayments created by the long barrier beaches of Long Island's south shore. Both bays are shallow, three to six feet deep, with a firm sand bottom that once supported a thriving hard-shell clam and oyster fishery. In 1910, Long Island Sound and the south shore bay oysters and clam business brought in $50 million annually; by 1960, it had dwindled to $15 million. In the early 1950s, dense blooms (over a million organisms per milliliter) of the alga *Nannochloris atomus* occurred yearly on the clam beds. It displaced their natural food, clogged their ciliary tracts, and proved to be directly toxic. That problem has since been alleviated by setting the pens back away from the water, but both bays now suffer the effects of a burgeoning human population in the area. Sewage effluent and non-point source runoff from suburban lawns have raised nitrogen loadings (hence algal blooms) high enough to deplete the clam beds.

Pathogens

Sewage and agricultural and street runoff bring high bacteria loads into our estuaries. Shellfish capture and accumulate the common intestinal bacteria of the higher animals and man. Although harmless in themselves, their presence opens the possibility of other, more virulent forms of disease-causing virus and bacteria lurking nearby. Whether these have been a source of infection for marine animals is not known. They have often been implicated in outbreaks of human ailments. Periodic outbreaks of gastroenteritis and hepatitis have been traced to contaminated shellfish, often illegally harvested from "closed" waters.

FOUR

A COASTAL SURVEY

The threats of pollution are not uniformly distributed along the coastline of the United States. With a few exceptions, they are concentrated in heavily settled and industrial regions. However, no coastal region is without some evidence of pollution.

The scope of the problem nationwide is just beginning to become apparent. Unfortunately, the diversity of pollutants, their uneven dispersal, and their little-known biological consequences, coupled with the multiplicity of marine ecosystems, have left us far short of a detailed picture of our national problem. In many unstudied coastal areas we do not even possess a broadbrush view.

Coastal waters have not been explored and charted as we have mapped our lands. There are no marine equivalents of the geological and agricultural maps that exist for nearly every square mile of U.S. land. No wetland atlases are available from the Government Printing Office or the Department of the Interior.

The information that exists is piecemeal. States maintain

shoreline descriptions, with information on shellfish beds and wetland zones. Still, the information on the impacts of wastes on these subsystems is scarce, scattered, and in constant flux.

All shorelines suffer at least minor pollution problems that are worth noting but pale in comparison to major examples of serious degradation. Even the major examples make up too long a list to cover adequately in a short space. But a brief coastline tour of our estuaries, stopping here and there to look a little deeper at a unique problem or an example of a widespread ill, can help one get a grasp of the overall dilemma.

NEW ENGLAND

The New England coastline consists of subsiding rocky headlands slowly being invaded by the sea. Narrow muddy and sandy embayments line its tidal rivers. Few coastlines are thought to be as pristine as those of Maine and New Hampshire. They have only a handful of small industrial coastal cities, located along rivers and bays that are flushed by swift currents and high tides.

Yet the harbor sediments of Maine's Penobscot Bay, Saco Bay, and Great Bay contain all the toxic metals—chromium, copper, lead, the lot—at about the same concentrations as are to be found in the more heavily industrialized areas to the south. The bottom sediments also hold polyaromatic hydrocarbons, not from oil spills but from fallout from wood fires.

Wood-stove smoke in the United States yearly emits almost as much airborne particulate matter as do all the nation's coal-fired power plants, more carbon monoxide than all American industry, and 40 percent of all polycyclic organic materials.

Dioxin, a by-product of paper manufacture, has been

found in fish from the Androscoggin River. Pulp-mill wastes plague rivers and streams of the East Coast from the northeast to the south. Pulp-mill operations are of two types—the sulfite process and the Kraft process. In the sulfite process, wood chips are cooked under pressure in a solution containing sulfur dioxide, which dissolves the lignin holding the cellulosic fibers together. The waste is a coffee-brown fluid with a pH of 2-4 (acidic). By contrast, the Kraft process uses an alkaline cooking liquor. Both effluents have a high BOD loading, spent sulfite liquor more so than Kraft waste. Both wastes drastically change the pH of the receiving water and render it turbid. Kraft waste is more toxic than sulfite liquor because it contains sulfides, mercaptans, and resin soaps. Continuous discharge of either waste soon coats the bottom with a sludge that kills off bottom life. Both kinds of waste in low concentrations degrade adult oysters by inhibiting their feeding. The eggs and larvae of oysters and clams are very sensitive to sulfite waste. Continuous exposure to low levels of these wastes will quickly denude an area of common New England invertebrates: barnacles, mussels, hydroids, and more.

If it comes to choosing the filthiest waters in the United States, Boston Harbor will be a leading contender. Forty-seven square miles of shallow bays and tidal estuaries, it receives the drainage of three rivers: the Charles, the Neponset and the Mystic. These three rivers, no gems of purity themselves, contribute only 35 percent of the annual input into the harbor; the rest is sewage and much of that is primary waste.

Harbor tides flush into Massachusetts Bay. Surprisingly, those currents are strong enough to keep some of the harbor open to shellfishing. About half of the four thousand acres of clam beds were open in 1985.

Boston's sewage problem reached scandalous proportions in the early 1980s. According to the Massachusetts Office of

Coastal Zone Management, the Metropolitan District Commission, the largest (85% by volume) of the thirteen districts that dump into the harbor, "has been a case study of management at its worst." They discharge primary waste, consistently fail to comply with state and federal standards for effluent quality, they dispose of sludge by pumping it out with the waste on outgoing tides, and their outfalls foul public beaches, shellfish beds, and lobstering grounds within the estuary.

The shoreline mess, along with well-publicized news about the growing levels of contaminants in shellfish (especially lobster), prompted the state to revamp the commission under a newly formed (1985) Massachusetts Water Resource Authority.

Harbor sediments in Boston are not unlike those of other polluted harbors: the usual nasty assortment of metals, hydrocarbons, and PCBs. In copper and lead, it is exceeded only by the Hudson-Raritan Bay complex in New York and New Jersey. For polyaromatic hydrocarbons (PAH), Boston Harbor's sediments contain some of the highest levels ever recorded.

"Polyaromatic hydrocarbons" is a generic term for a host of organic compounds, many of them cancer-causing agents. Their threat to fish and man is long term. In fish it manifests itself as abnormal development, deformities, impaired growth, genetic damage, and tumors.

Fish from Boston Harbor show high incidences of cancerous lesions. Of a 1984 sample of bottom fish (mainly winter flounder), 20 percent had liver lesions. In a March 1985 survey, 64 percent had lesions and 42 percent suffered liver tumors. No estuary on the East Coast has a higher incidence of carcinoma in its catches. The cause is unknown, but PAHs are strongly suspected.

New England ports handled 590 million pounds of fishery products in 1985, worth $419 million. Leading the list was

New Bedford, the leader not only for New England but for the United States, doing a $103 million business that year. It is also the most severely polluted harbor in New England, and its waters, both the Inner and Outer Harbors, are closed to all finfishing, shellfishing, and lobstering.

New Bedford has the highest concentrations of PCBs in the United States in its harbor-bottom sediments. Sewage pollution keeps the Outer Harbor and nearby Clark's Cove beds closed. The city discharges 23 million gallons of minimally treated (less than primary) waste each day. Its single treatment plant has never met the standards for primary treatment since it was built in 1973. Because the region has a combined sewer system, during every rainstorm 38 outlets drain raw sewage directly into the harbor and nearby coves.

In the mid 1970s, the discovery that harbor sediments were saturated with PCBs led to a ban on all fishing. The yearly loss because of these fishing and shellfish closures is estimated to be $22 million. Other New Bedford harbor problems include high fecal coliform counts, and black gill disease and shell disease that afflict over fifty percent of the lobsters in the closed waters.

West of Massachusetts lies Rhode Island, dominated by Narragansett Bay. The wedge-shaped bay is the center of the state's industry and recreation. The bay is fairly narrow, deep, quite salty (25–30 ppt) and clear. It has few sea-grass beds or kelps. Its productivity is based on phytoplankton.

At one time, Narrangansett Bay had a thriving oyster fishery. It is gone now, but productive clam beds remain. The bay is an important nursery for winter flounder and sand lance, and an occasional hot spot for menhaden, attracting commercial porgy boats from as far away as North Carolina.

The problems of the bay in the early 1980s were like those of hundreds of coastal cities along our seaboard, re-

sulting mainly from poorly designed and maintained municipal waste systems. The city of Providence, like so many others,
had opted for a combined sewer overflow system (CSO). The
piping layout for such a system is arranged so that dry weather
flow dumps into an interceptor through slots. When it rains,
the slots cannot catch the total flow, which then heads down
an outfall pipe to the sea. At the end of the outfall pipe is a
tide gate: a vertically hung iron flapper valve that shuts at
high tide from the back pressure of the rising seawater and
is free to swing open at low tide from the pressure of outfall
runoff.

If the slots are not periodically cleaned out (which they
usually are not), they soon jam up with street litter. This
prevents sewage from entering the interceptor even in dry
weather and guarantees a steady flow of raw sewage into the
outfall. Similarly, if the tide gate jams open, seawater will
back up the outfall during high tide and enter the interceptor.
Seawater is cold as well as salty and quickly reduces the efficiency of a secondary treatment plant.

A number of municipalities dump treated waste into the
bay. Some operate well; others, poorly. "Save the Bay," a
Rhode Island conservation group with a growing national reputation for shaping public opinion, conducts a quarterly survey
of regional sewage plant performance. They call it "The Good,
the Bad and the Ugly." Plants are rated Good if they operate
in compliance for ten to twelve months of the year; "Not-So-
Bad", six to nine months; Bad, one to five months, and Ugly,
not at all.

Out of twenty-four plants in the region, fifteen are now
in the Good category and seven in the Bad and Ugly groups.
Their reports are widely publicized by the local press and
have exerted considerable pressure on local politicians to improve the situation.

The University of Rhode Island has a major School of

Oceanography, which uses the bay extensively for study and research. One such study of urban runoff of oil pointed up the influence of major highways as a source of petroleum hydrocarbons and PAHs. Only a seaside industrial area that included an oil terminal proved to be a heavier contributor.

Citizen pressure and a responsive state government are improving the health of the bay. Oil and grease loadings are down significantly, mainly because the industries responsible have either left the state or closed down. Better maintenance of CSOs and treatment plants has substantially reduced the

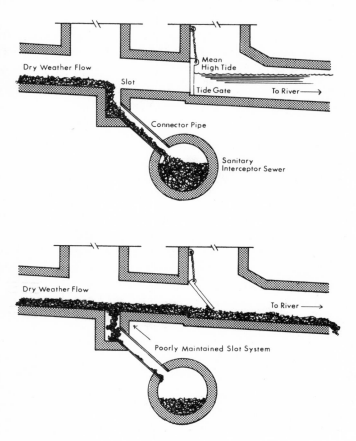

amounts of untreated and poorly treated waste reaching the bay. The area still has substantial problems, however: Toxic waste illegally dumped by small electroplaters in the Providence area has yet to be staunched, and the rapid urbanization of the shoreline continues unabated.

MIDDLE ATLANTIC STATES

The eastern end of Connecticut and Long Island mark the southernmost extent of the Paleozoic rocks of New England and the beginning of a long, sandy coastline that stretches down the rest of the eastern seaboard to the middle of Florida, where the quartz sands are replaced by those derived from limestone.

Technically, all of Long Island Sound is an estuary. One hundred and ten miles in length, averaging fifteen miles wide, its eastern end is bounded by Block Island and a series of small islands that sit atop a submarine ridge. The western end narrows to the confluence of the East and Harlem rivers, tidal straits that connect the Sound with the Hudson River and lower New York Bay.

Ten percent of the nation's population live within fifty miles of the Sound. It harbors the largest pleasure-boat fleet in the world: over 125,000 vessels. Each year more than a million fishermen spend 12 million days on the Sound (1975). Its small commercial fishery includes occasional menhaden, winter flounder, and small harvests of clams, oysters, and lobster. Finfish populations are low compared to coastal waters. This has been attributed to its muddy bottoms, where there is not much forage compared to what is found in adjacent waters.

The water at the eastern end of the Sound is relatively clean, but westward the sewage burden of metropolitan New

York and Connecticut overload the water with nutrients. East River treatment plants dump sixty tons of nitrogen into it every day. All other plants at the western end dump another thirty tons daily.

Occasional episodes of bottom water anoxia have hurt the lobster fishery in the western end. Thermal pollution, consisting of warm water from electrical generating plants, is suspected of being associated with lymphocystis, a viral disease of fish. Striped bass from central Long Island Sound, around the Northport effluent, have a high incidence of infection. Seventeen generating plants on the Sound use 2.1 billion gallons of cooling water every day. Poor water quality has closed much of the 125,000 acres of shellfish grounds around the Sound.

Daily nitrogen loadings in the western quarter are about three times higher than those of Narragansett or Delaware Bay. Lower New York Bay, by contrast, receives a daily loading (half a kilogram per square meter per day), ten times higher than that of the Sound, more than thirty times that of the average East-Coast estuary.

Lower New York Bay and the adjacent Raritan Bay are battered by a 650-ton-a-day BOD loading. Surprisingly, except for low summer episodes, the dissolved oxygen content of the water at the mouth of the bays averages near saturation levels. Phytoplankton growth is a major factor in maintaining oxygen levels in the upper waters. Seabed oxygen levels are low, the consequence of decaying organic matter in the sediments.

Summer phytoplankton blooms recur regularly, averaging one to two weeks in duration. These blooms only occasionally turn destructive. Phytoflagellate blooms have killed fish and, where they touch the shore, have caused respiratory problems among swimmers.

As a result of discharges from two General Electric plants

near Albany between 1950 and 1976, sediments in parts of the Hudson are laced with PCBs. The lower Hudson contains PCBs in concentrations exceeding ten parts per million, one to two orders of magnitude higher than that present in any other large river system or estuary on the East Coast.

PCBs are found in the larvae and adults of the striped bass, bluefish, American eel, white perch, and catfish, often accumulating to levels well above the 2 ppm allowable marketing limit set by the Food and Drug Administration. The sale of striped bass and eel is prohibited in New York and Rhode Island because of high PCB levels.

Small amounts of PCBs have been found in a long list of edible species from the metropolitan area: fluke, winter flounder, anchovy, whiting, tautog, hard clam, blue mussel, oyster, blue crab, and lobster. PCBs have also been detected in ospreys' eggs.

Lying in the path of the Hudson, Passaic, and Raritan rivers, Raritan Bay is a repository for all the common industrial and sanitary wastes that urban society has to offer: toxic metals, hydrocarbons, sewage by-products, and more.

Except for ship channels, the bay is shallow. Its bottom is dominated by bristle-worms. However, it is the summer residence for a long list of recreational and commercial fishes: summer and winter flounder, red hake, scup, weakfish, butterfish, and bluefish. Some of these species suffer a fin erosion ailment called fin rot. It especially afflicts winter flounder and bluefish. Local invertebrates (crab, lobster, and shrimp) suffer a shell disease caused by an unknown component of sewage sludge.

Considering its location and the amount of business done along its shores, Delaware Bay could be doing a lot worse than it is. Over seven million people live and dump upstream, overnutrifying its waters and loading its sediments with metals, oils, and farm runoff.

Seventy percent of all the oil arriving on the eastern seaboard is delivered through the Newark-Wilmington area. The lower bay, on both the New Jersey and Delaware sides, is lined with farms. Upstream lies the city of Philadelphia. Thus its shellfish show bioaccumulation of metals, organo-pesticides, and hydrocarbons. Oyster yields are declining, as is the fishing.

As if that were not threat enough, there remains the long-term fate of the river flow. The upper reaches of the Delaware River are pristine and lie in New York State, where the de-mands for drinking water for metropolitan New York increase yearly.

Damming the river at Tock's Island and diverting more water to the New York reservoir system has been proposed to fill this demand. Downstream the effect of this would be saltwater intrusion upriver. Salinity profiles of the bay during times of flood (March 1936) and drought (November 1930) showed vast shifts in both the bay and the river. A drought in 1962–1965 reinforced the folly of reducing the water flow further. Salty water made its way as far upstream as Phila-delphia, jeopardizing the fresh water supply for hundreds of thousands of people.

Chesapeake Bay is the largest estuary on the East Coast, covering an area of 4,400 square miles and draining a wa-tershed of 64,000 square miles. Norfolk sits at its mouth and Baltimore near its northernmost end. Both of its eastern and western shorelines are heavily indented with coves, wetlands, and river systems: the James, the York, the Rappahannock, the Potomac and the Susquehanna, which create a total shore-line seven thousand miles long.

The bounty of the bay is legendary. Writing about his boyhood in the 1880's, H. L. Mencken referred to it as "the immense protein factory" and noted, "Any poor man could go down to the banks of the river, armed with no more than a length of stout cord, a home-made net on a pole, and a chunk of cat's meat, and come home in a couple of hours with enough crabs to feed his family for two days."

Today the bay is in trouble. In 1890, the bay produced 7 million bushels of oysters a year. In 1988, yield predictions go as low as less than a million, maybe far less. Blue-crab catches have fallen drastically. Stocks of Atlantic striped bass for whom the bay is a major spawning ground and nursery collapsed, but, through hatchery efforts, they are now making a comeback.

The demise of this rich ecosystem can be traced to over enrichment, toxic influxes, increasing turbidity, and the al most complete collapse of its once abundant sea grasses.

The nitrogen burden flows in through the river systems The James, the Potomac, and the Susquehanna contribute sewage; the Rappahannock and the York and the tributaries of the eastern shore add agricultural runoff. Most of the de veloped land in the Chesapeake drainage basin is farmed mainly in corn and soybeans. These farms have more than doubled their use of fertilizer since 1964.

The Environmental Defense Fund claims that acid rain has become a heavy nitrogen contributor to the bay and indeed, to all the other estuaries of the East Coast. The fund estimates that 34 percent of the bay nitrogen comes from fertilizer runoff, 25 percent from acid rain, 23 percent from sewage, and 18 percent from animal waste.

Reports of brief periods of anoxia in the bottom water of the bay go back to the 1930s. Over the years, intervals of oxygen depletion have lasted longer and become more geo graphically widespread, lasting from May to September and extending from Baltimore to the mouth of the Potomac.

In 1984, a catastrophic combination of natural events held the oxygen-deficient bottom water in place long enough for it to go anaerobic, producing hydrogen sulfide that killed nearly everything on the bay bottom deeper than eighteen feet.

Over the last fifteen years, the bay has lost almost all its submerged aquatic vegetation. The decline in the sea grass *Zostera marina* exceeds that of the pandemic of the 1930s. Unlike the 1930s losses, which were caused by a wasting disease, the grasses are not growing back. Turbidity, caused by erosion upstream, plays a role; herbicides may also be involved.

Pinning biotic degradation to toxic dumping is seldom easy. The illegal pumping of Kepone-laden waste into the James River at Hopewell, Virginia, for over eleven years (it finally ceased in the mid-1970s) is still poisoning fish and shellfish downstream. Kepone is a chlorinated hydrocarbon, as is Mirex, a pesticide widely used in the South to control fire ants. Heptachlor is similar, and a fish killer, as are dieldrin and endrin. The Food and Drug Administration is authorized to block the sale of shellfish containing more than 0.1 to 0.3 ppm of these pesticides. (The extent of inspection is another matter.)

The harbor at Norfolk, a major naval base, reflects another environmental hazard, tributyl tin (TBT). TBT is a biocide used in anti-fouling hull paints. TBT paint lasts longer and is far more toxic than copper-based paints. Extremely low concentrations (parts per trillion) have toxic effects on oyster and clam larvae. Tributyl tin can be bioaccumulated by some shellfish and reach levels harmful to man. Nearly 50 percent of all the recreational boats in the bay, as well as many large commercial and military ships, use TBT. EPA may restrict its use to deep-water vessels only.

Compared to the bays and sounds of the northeast, the coastal waters of North and South Carolina and Georgia are only moderately polluted. Nutrient overloading from sewage, fertilizer runoff, and pulp-mill and phosphate-mining waste have eutrophied local waters, and high coliform counts have closed shellfish beds.

Albemarle and Pamlico sounds, those shallow waters lying behind the barrier beaches of Cape Hatteras and the Outer Banks, are beginning to feel these combined effects: finfish landings are down, shrimp catches are falling, and algal blooms are killing fish.

From North Carolina through Florida, commercially important deposits of phosphate nodules and rock are found on land and close to the shore. To extract phosphate, the rock is pulverized, then concentrated by a flotation process. The residue, fine clays, and ultra-small particles of phosphate-bearing particles are held in basins called slime ponds. These particles slowly settle out but retain enough water to remain semifluid for years. If the impoundment walls fail, the settlings easily escape and resuspend in moving water to later resettle as a clogging sediment.

Just such a dike break occurred in 1967 along the Peace River, which discharges into Charlotte Harbor, Florida. It killed a million fish outright and coated the banks and bottom of the river with a thick layer of slime that persisted for years.

FLORIDA AND THE GULF STATES

Tampa Bay also receives phosphate waste from an upriver operation. Combined with sewage enrichment and decaying water hyacinths, it has created eutrophication seldom equaled in the United States.

The Florida Keys, a 130-mile chain of low-lying islands lined to seaward by shallow coral reefs and to bayward by mangrove islets, is a fragile ecosystem now overrun by development. Linked to the mainland by a single highway, its population will soon exceed one million. The impact of this population, as well as the sediment drifts from the Miami area, have degraded its reefs and inshore water quality.

The future of the Everglades, that sea of saw grass and cypress hammocks, seawardly lined with dense mangrove stands, rests entirely on man's control of its supply of fresh water. The construction of the Caloosahatchee Canal in 1882 first started the alteration of its hydrology. Canals and dikes in the 1920s dropped water levels in the Everglades by ten feet. By the 1940s the state awakened to the potential problems of saltwater intrusion. More construction ensued, impounding fresh water to create a barrier against the sea. These changes lowered the freshwater input to Florida Bay, a major shrimping ground, decreasing its productivity.

From Mobile Bay at the western edge of Alabama to the eastern end of Louisiana lies a long, shallow embayment known as the Mississippi Sound. Shoreward, it is lined with marshes, bayous, and coves. Offshore, it is bordered by chains of small islands. The Sound is a major nursery ground for fish and the site of a substantial commercial fishery producing 470 million pounds a year, mainly menhaden and shrimp.

The Sound is suffering the now familiar legacy of progress: metals and organics in its sediments, eutrophication, hypoxia, and pathogen contamination. Especially troubling are the high concentrations of chlorinated hydrocarbons emanating from the Pascagoula River system and Biloxi Bay.

The wetlands of the Mississippi River Delta are one of the earth's most productive ecosystems. Some 40 percent of the coastal marshes in the United States are located along Louisiana's coastline, where Old Man River meets the sea.

The delta's commercial fisheries landings exceeded 1.6 billion pounds in 1986, over 25 percent of the entire U.S. catch. The wetlands are also the winter home for nearly four million waterfowl, more than two-thirds of the birds that use the Mississippi flyway.

But this great wetland community is losing its swamps and marshes to the Gulf of Mexico. Coastal lowlands are eroding into open water at the rate of fifty square miles a year, the result of dredged navigation channels, boating canals, flood control levees, and sinking land.

Early canals were dug for boat traffic and smaller ones called "trainasses" were cut by trappers to get between bayous. After World War II, new canals were dug to get to and from oil- and gas-drilling sites. As the canals were created, so were spoil banks along their borders. The net effect of the canals, the huge levees built to contain the river, and the dredging for navigation and flood control has been to isolate the delta wetlands from the river sediments and fresh water that sustain them. River sediments now go straight to sea, lost to the deeper waters of the gulf.

Between 1956 and 1978, 560,000 acres of marsh have been lost. If nothing is done about it, another million acres, an area the size of Rhode Island, will vanish by the year 2040. The intruding gulf water is not only washing away the former alluvial deposits of the river but is also killing off plants not tolerant to salt water.

The coastline waters of Louisiana and large areas of the Gulf of Mexico have also recently suffered extensive bouts of hypoxia. The causes are not understood, but both the Mississippi and Atchafalaya rivers are laden with nutrients and pollutants.

Westward lies Texas, whose arid gulf borders are lined with long, narrow barrier islands that enclose very salty lagoons. This harsh environment of algal mats and sea grasses supports an ecosystem easily destroyed by sewage pollution. In many areas, dredge spoil has smothered productive oyster beds and cut off circulation to grass flats, killing them.

Houston ship channel, Corpus Christi Harbor, and Galveston Bay are examples of multiple-stressed waterways, much like the Kill Van Kull and Newark Bay in New York and New Jersey. The sediment in these waterways contain so many contaminants they have been compared to the primary sludge of an urban sewage treatment plant. Periodically, these channels must be dredged to keep them open to ship traffic, and the dilemma arises of what to do with those toxic sediments. Usually they are dumped at sea.

WESTERN STATES

Fifty-two percent of California's wetlands and estuaries have been destroyed by dredging and filling. Of the remainder, 81 percent have suffered moderate to severe damage. In southern California, 75 percent of the wetlands are gone and all major harbors are badly polluted.

The harbors of the port cities of San Diego, Los Angeles, and San Francisco have all been damaged by sewage and oil. Spills and oil-refinery waste seepage in Los Angeles–Long Beach harbor have eliminated sea-bottom life except for a single oil-tolerant bristle worm. Both the toxicity of the effluent and its oxygen-demand exacerbate its effects.

Oil effluent from Los Angeles refinery operations runs about 10 to 100 ppm, seemingly not much, but the volume of effluent is so great from a major refinery that several thousand gallons of oil a day can escape. Petrochemical plants have a similar impact, releasing an even wider array of toxins.

The San Francisco estuary has the dubious distinction of being more highly modified by the hand of man than any other estuary on our coasts. The bay lies at the mouth of the Sacramento and San Joaquin rivers, which carry 40 percent of California's runoff. The changes started with the 1848 gold rush. By 1900, salmon, sturgeon, and Dungeness crab were commercially fished out. All that is left today is a herring and anchovy fishery.

In 1869, live eastern oysters (*Crassostrea virginica*) were transplanted onto the bay's mud flats. They failed to take hold, but a number of invertebrate fellow travelers thrived including the oyster drill and the shipworm, which promptly infested all the wooden piers, pilings, bridges, and trestles in the area.

Striped bass were also introduced in 1879 and did quite well by themselves until 1915 when overfishing closed the commercial fishery. Today, hatchery-raised, they are a major sports fishery. However, bay bass suffer high incidences of lesions and skin tumors thought to be induced by contamination by organic chemicals.

In 1853, up in the Sierra Nevada, gold mining changed from pick and pan to hydraulic sluicing. By 1884, when sluicing was banned, tens of millions of tons of dirt and rock had washed downstream, choking creeks and rivers throughout the drainage basin. Much of this sand and mud eventually reached the bay, ruining the oyster beds and eliminating the salmon's spawning grounds.

Freshwater marshlands and tidal marshes were diked and drained, at first for farmland, later for saltworks and development. Of the original 8500 square miles of tidemarsh in 1848, less than 50 square miles remain today.

Agricultural demand for irrigation water, both north and south of the bay, has led to the impoundment and diversion of the Sacramento and San Joaquin rivers. Freshwater flow into the bay has been reduced by 60 percent. New water projects will reduce that to over 70 percent by the year 2000. This continual rise in the bay's salinity spells an end to organisms that need brackish water to survive.

The San Joaquin River flows through Central Valley, which is arid, heavily irrigated, heavily fertilized, and liberally laced with herbicides and pesticides. Irrigation leaches salts from the land. Irrigation runoff returns to the river and makes up 20 percent of its total flow to the bay.

The volume of agricultural wastewater is increasing. To dispose of it, a drainage channel was proposed to carry it to the bay. Partially built and operational by 1978, it carried 10 percent of the valley runoff into Kesterson Reservoir, a marshy wildlife refuge. Selenium, leached and concentrated from farmland soil, devastated aquatic bird life and vegetation in the refuge. That will soon be stopped and all the valley flow will go directly to the bay (with what effect remains to be seen).

At the northwestern extent of the continental United States lies Puget Sound, whose adjacent waters extend well up into British Columbia. They connect with the sea through the Strait of Juan de Fuca.

The Sound lies in an enormous basin between the Olympics, the Cascades, and the southern lowlands. Islands, promontories, and long, narrow, involuted inlets divide the Sound south of Admiralty Inlet into four subbasins bounded by sills and narrows. The basins are deep, 600 feet and more, and the boundary sills are 150 to 250 feet deep. Seven cities border its waters from Port Angeles at the strait to Olympia far to the south. Between lies Seattle.

The Sound is the centerpiece of Washington State; 56 percent of its residents visit its shores at least once a year. Tourists flock there to see its whales and seals and to fish. The combined commercial and recreational fish and shellfish harvest amounted to $74 million in 1984.

Commencement Bay, off Tacoma, has one of the most contaminated sediment beds in the country, so badly burdened by toxic metals and organics it is at Superfund site levels. Organic contaminants, PCBs and PAHs, have accumulated in the tissues of fish, crab, clams, and mussels. Liver abnormalities are common in ground fish, particularly the English sole. Similar sediment catastrophes also exist near other big cities of the Sound.

FIVE

COASTAL SEAS AND OPEN OCEAN

A glance at a map of the continental United States gives the erroneous impression that deep oceans extend undifferentiated right up to the shoreline. In fact, the waters of the East and Gulf coasts remain relatively shallow far to seaward. On the West Coast shelves do hug the shoreline and its underwater borders are not far removed from those shown on a conventional map.

These broad underwater plains, deeply gullied with basins separated by ridges in the northeast and low, sandy ridges and swales along the rest of the East Coast and the gulf, are among the world's most productive fishing areas. The banks and ledges of the Gulf of Maine and Massachusetts Bay— Georges, Stellwagen, Davis, Jeffreys—are legendary fishing grounds for cod, pollack, haddock, yellowtail flounder, hake, halibut, mackerel, herring, and a dozen other species.

Some of these waters have suffered seriously from pollution. The cold waters of the north are still relatively clean, but other areas are no longer so pristine. Long coastal stretches off Louisiana are annually swept by oxygen depletion. In the

New York Bight fish and shellfish suffer fin rot and shell disease, and those waters suffer occasional bouts of low oxygen.

In some places, the signs of pollution extend well out to the edge of the continental shelf. Pelagic fishes turn up loaded with hydrocarbons and PCBs in their tissues. Bottom sediments well offshore contain traces of sewage-related material. Occasional events, partly triggered by natural causes but exacerbated by pollutants, destroy sea-bottom life over huge areas.

The open ocean is not particularly productive, especially in warmer seas. It is nitrogen- and phosphorous-poor, and the growth of phytoplankton is limited. Without land runoff, the food chain must depend on getting its inorganic nutrients either from the upwelling of nutrient-rich cold water, which occurs only in specific regions, or from the recycling of animal waste products and decaying life. However, nutrients are continually lost to the deeps.

On the shelves, it is an entirely different matter. Land runoff carries with it suspended solids, particulate matter, and dissolved inorganic salts that can support life at a far brisker pace than can open ocean. These nutrients are not easily lost from the cycle. Debris sinking to the bottom supports a healthy sea-bottom community which, in turn, is within easy reach of the swimming foragers.

It is hard to assess the health of the oceans. Fishery statistics mean little or nothing because overfishing has been rampant during the last twenty to thirty years. In the late 1950s, huge vessels began fishing the Grand Banks; these factory trawlers fished for months at a time, twenty-four hours a day, seven days a week. Equipped with state-of-the-art electronics and huge mid-water and bottom trawl nets, they swept the waters clean. In 1976, the United States unilaterally extended its fishery jurisdiction two hundred miles to sea to stop this pillage. Other nations followed suit. The trawlers went elsewhere, but the stocks of many fish species were severely set back and have taken years to recover.

The vagaries of nature play a leading role in the success of a year-class of fish. The presence of predators at a critical time during the early stages after spawning can decimate the ranks of a new year-class. The voracious arrowworm can cut down herring larvae at an astounding rate, and a sea teeming with them bodes ill for the survival chances of a herring set. Species have disappeared for years from what has been presumed to be natural causes or overfishing, only to reappear and prosper once again.

We can, however, follow the fates and effects of man-made contaminants at sea. Finfish from the Georges Bank, the mid-Atlantic bight, and the Gulf of Mexico have trace quantities of PCBs and petroleum hydrocarbons in their tissues. Some have accumulated levels close to federal action limits. Silver hake taken between Nova Scotia and Delaware have shown PCB levels from 0.1 to 0.5 ppm and PAHs from 1 to 100 ppm. PCB levels in bluefish, striped bass, white perch, and eels taken off the New Jersey coastline have exceeded federal action limits of 5 ppm.

PCBs can be spread through sediment distribution, in the water currents, or by air pollution. Concentrations in fish diminish the farther they are caught from centers of pollution, like the mouth of the Hudson River, suggesting that water currents are the main method of transport.

What do sublethal levels do to the well-being of fish and their reproductive success? Not much is known, but PCBs in striped bass and white perch accumulate more in reproductive cells than in muscle or other body tissues.

Particulates and organic debris in water are especially high along our coasts. Effluents from estuaries tend to travel alongshore rather than go directly to sea. At sea, suspended matter in the water contains mostly soot, fly ash, and cellulose, suggesting that its source is fallout from polluted air.

Seabound particulates from ocean dumping, dredge spoil, sewage outfalls, and other sources carry, distribute, and settle out pollutants over vast areas. You cannot track sewage by fecal coliform bacteria counts too far from its source once it enters salt water, but you can follow the track of coprostanol, a compound manufactured in the mammalian gut and relatively persistent in sea sediments.

New York City and some New Jersey sewage sludge used to be dumped at sea at a site 12 miles offshore, which has since been moved to a new site 106 miles offshore. Sewage sludge dumped at the old twelve-mile site off New York City contained about 10 ppm coprostanol. Concentrations as high as 0.1 ppm were found in sediments as far as thirty-five miles from the dump site. Similarly, metals and hydrocarbons had been carried long distances from their point of first discharge.

A number of seaside municipalities directly discharge their waste into coastal waters. The law mandates that their discharges be secondary waste. By a legal exception California outfalls are allowed to pump primary waste directly to sea. The area surrounding their outfalls is seriously degraded, but as Pacific currents sweep the suspended matter away into deep water, damage is not obvious farther away. California has argued, given its proximity to very deep water and swift currents, that the answer is to extend its pipelines, not upgrade sewage treatment.

SLUDGE

Slurry-like settlings from primary and secondary waste treatment can be dewatered and incinerated, landfilled, used as fertilizer, or dumped at sea. About 10 percent of all sewage sludge that is produced in the United States is dumped into the ocean from ships or through pipelines.

Only Boston and Los Angeles County discharge sludge by pipeline. In 1980, southern California waste-treatment plants were disposing of 107 thousand dry tons annually this way. Both will phase out this practice, although Boston will probably continue until the mid-1990s. However, Orange County, California, has proposed an eight-mile-long pipeline to the edge of the continental shelf through which it wants to discharge relatively uncontaminated sludge. The issue remains unresolved.

After the 1977 amendments to the Clean Water Act banned the dumping of sludge that would "unreasonably degrade" the ocean, more than a hundred small municipalities stopped sea dumping. These accounted for only 5 percent of all dumped sludge. A 1981 court decision (*City of New York* v. *EPA*) allowed nine sewage authorities in New York and New Jersey to continue the dumping. They had used the 12-mile sewage sludge dump site in the New York Bight since 1924. This site has been phased out and the sludge is now hauled to the 106-mile, deepwater municipal site.

The 12-mile site was a central target among environmentalists. Studies on the distribution of waste, metals, coliforms, PCBs, PAHs, and other contaminants showed that they dispersed away from the site, entered the surrounding ecosystem, and created a "dead sea" there and nearby.

Beneath the dump site the bottom had the consistency of black mayonnaise. The only things that lived in it were bacteria and a few species of polychaetes including *Capitella capitata*, the sea-going sludge worm.

Significant numbers of fecal coliform bacteria from those sediments and the surrounding water turned out to be resistant to common antibiotics, sulfa, and heavy metals. Drug-resistant *E. coli* persisted in the sediments for over one month.

Pathogenic amoebae, *Acanthamoeba*, were common in bottom samples. This and the presence of other bacteria and viruses gave rise to the fear that this area could be a human health hazard.

Bacteria found in the sludge has been implicated in fin rot disease, a skin necrosis that has hit bluefish, summer flounder and weakfish in that area. The disease proliferated widely in 1967 and has persisted ever since.

Although it appears open to the ocean, New York Bight water, which is bordered by Long Island and New Jersey landward and the edge of the continental shelf seaward, circulates slowly, remaining in the Bight for a long period of time. Except in winter the water is stratified horizontally. Summer top water can reach 75°F while the bottom water remains cold (54°F or less). Continual onshore winds from the southwest and a paucity of storms that otherwise might move the water can keep this stratification intact, setting the stage for environmental disaster. Loaded with nutrients flowing from the Hudson-Raritan estuary and from sludge and spoil sites, summer algae blooms in the top water soon collapse, showering the colder, trapped bottom water with organic oxygen-consuming debris, creating hypoxic conditions.

Although hypoxic events had been noticed since 1949, on July 3, 1976, while the tall ships entered New York harbor for the bicentennial celebration, the largest bottom kill in the history of the Bight started. Three thousand square miles of ocean bottom began to die. Bottom mortalities were astronomical. The surf clam losses alone were 140,000 tons of meat, 70 percent of the offshore population. Heavy losses also occurred among ocean quahog, sea scallops and lobster. Although the incident was attributed to natural causes at the time, government reports later indicted the flow of waste into the Bight as a contributing factor.

Moving the metropolitan New York sludge dump site

farther offshore has not ended the controversy with environmentalists or fishermen. The deep water site, on the continental slope, is fished for lobster, tilefish, and squid. Lobstermen
have reported shell disease in their catches, the first time this
affliction has been seen so far to sea, but federal marine scientists do not believe it is related to dumping.

Alternatives for sludge disposal depend on the type of
solids they contain. Primary waste solids are 75 percent organic matter, of which a third is grease. This is obnoxious
stuff and not suited for agriculture. Secondary waste, anaerobically digested, contains 60 percent organic matter of which
5 percent is grease and can be used on farm or forest. However, sludge laden with toxic metals used on cropland results
in high metal uptake in plants, a drawback to its application.

Dried sludge can be incinerated using methods that minimize air pollution. The ash, essentially free of organic matter,
does retain and concentrate heavy metals and presents a landfill problem.

As the United States upgrades from primary to secondary
treatment, considerably more sludge will be generated. The
total amount has doubled in the last ten years, and, by the
year 2000, so estimates a congressional report, yearly output
could rise to 10 million dry metric tons.

According to the Office of Technology Assessment, "The
management of sludge is controlled by a patchwork of federal,
state and local regulations. . . . Arrangements among municipalities, counties, states and EPA regions are highly site-
specific and complex, and often highly politicized. Management also is complicated by a lack of comprehensive disposal
standards, changing economic conditions, public opposition,
and a relative lack of promotion of the idea that sludge can
be used as a beneficial resource."

As the coastal population grows (by 1990, 70 percent of
the U.S. population will live within an hour's drive of the

coast) the demand to use the ocean for disposal will grow. Should sludge dumping be allowed in marine waters? It depends on the sludge. Uncontaminated material "could probably be dumped in open ocean without causing severe impacts," according to the OTA, but that would mean throwing away a valuable resource that should be used on land. Contaminated sludge, the bulk of what is destined for marine waters, "argues in favor of a policy that calls for restricting (at least to some degree) the dumping of sludge. . . . If marine dumping of sludge is to continue, it seems prudent to use a dispersal strategy (e.g., dumping in well-mixed and deep waters) and to minimize the presence of metals, organic chemicals and pathogens. This strategy would clearly preclude dumping in estuarine and poorly-mixed coastal waters." Some environmentalists would object to *any* ocean dumping.

DREDGE SPOIL

Harbors, canals, and channels continually silt in and need periodic cleaning. Eighty to 90 percent of all waste material dumped in marine waters originates from dredging—180 million wet tons a year. As of January, 1987, about 126 dredge disposal sites were located in coastal waters and open ocean. Although half these sites are used each year, 95 percent of the spoil is dumped at two dozen locations.

Sixty-five percent of all spoil is dumped in estuaries; the remainder, offshore. Offshore and coastal dumping has varied widely over the last decade—between 35 and 120 million wet tons a year—depending on the activity in the lower Mississippi River, one of the country's siltiest water courses.

Dredge-spoil composition varies from clean sand, gravel, or silt to cesspoollike muck loaded with heavy metals, hydrocarbons, organics, and nitrogen. About 3 percent, or 7 million tons, is highly contaminated. Metropolitan New York spoil is among the worst, and its harbor spoils are a far greater source of organic carbon and nitrogen than sludge. Dredge spoil from New York, dumped in the bight not far from the old sludge site, adds 540 metric tons of carbon a day compared to 110 tons for sludge. Similarly, spoil adds 63 metric tons a day of nitrogen; sludge, 17 tons.

Some bottoms are so foul, like that of New York's Kill Van Kull, that dredging has been delayed because of arguments over its disposal. None of the choices now available are ideal: Move it to an upland fill (and worry about leaching), build a containment island (and worry about leaking), or dump it at sea and cap it with uncontaminated sediment (and worry about its moving). All three methods are now in use in the United States.

Uncontaminated dredged material, improperly handled, can wipe out bottom communities. A quahog fishery near Narragansett Bay was smothered in 1969–70 by several million tons of spoil. In earlier years, the Raritan oyster beds were similarly ruined by hydraulic dredging. As they saw their livelihoods vanish, local oystermen took to winging a few live rounds at the dredges but soon lost out to progress and the law.

The biological impacts of redistribution of metals, PCBs, hydrocarbons, and toxins are hard to assess, but moving material from a point of natural collection to open water hardly helps contain it.

The quantities of pollutants moved to sea this way are substantial. Boston Harbor spoil is dumped in Massachusetts Bay and loads the bay with 4400 tons of PCBs, 2100 tons of chlorinated hydrocarbons, 1760 tons of petroleum hydrocarbons, and 88 tons of cadmium each year.

In estuarine areas, federal and state agencies have favored disposing of spoil by dumping it inland in containments rather than moving it to another location within the estuary. However, doing this increases the risk of resolubilizing toxic metals bound to these sediments by exposing them to air and redistributing them through leaching and runoff.

OIL

Nearly 70 percent of the world's oil output travels by sea. Crude production in 1988 will be between 1.5 and 2 billion metric tons. A consensus of estimates for spillage from all causes ranges between 1 and 10 million tons annually.

Major accidents and blowouts make the headlines. The Torrey Canyon, Argo Merchant, the Amoco Cadiz, and the Ixtoc I blowout are well known. However, the major sources of petroleum pollution in the sea come from river runoff, coastal waste, natural seeps, and ship operations:

Annual Sources of Crude Oil and Petroleum Products in the Ocean

SOURCE	AMOUNT (thousands of tons, metric)
Tanker operations	889
Other ship operations	302
Vessel accidents	173
Offshore production	118
Natural seepage*	600
Nonmarine operations and accidents**	1,974
	4,056

Source: Travers and Luney, 1976.
*This number may be seriously overestimated—Blumer, 1972.
**1,035 kt are from waste automobile oil!

In mid-ocean, the overt signs of oil are floating blobs of tar, from a fraction of an inch in diameter to baseball size.

Oceanographic expeditions to the Sargasso Sea were among the first ships to notice them. In 1968, the *R/V Chain*, towing surface nets, came up with nets so fouled that towing was stopped. Crewmen estimated there was three times as much tar as Sargassum weed.

These lumps collect along windrows on the water's surface and intermingle with kelps and Sargassum weed. As they gain density from debris accumulation and loss of volatiles, they sink or beach. Bermuda's gray limestone rocks are speckled with tar spots and those that wash up into the sands easily transfer to the soles of bare feet. Their source is tanker cleaning. The Bermuda Biological Station estimated that in 1980 100 tons of tar a year were washing ashore.

The fate of spilled oil depends on the nature of the crude. Number 6 heavy crude barely floats and is very viscous. Number 2 heating oil (Diesel fuel) and light crude, #4, are lighter and less viscous, spreading easily across the sea's surface.

The light oils contain more aromatic hydrocarbons and lower boiling alkanes than does heavy crude, which contains a high percentage of biodegradable waxes. Low boilers induce anesthesia and narcosis among fish and invertebrates and, at higher concentrations, death. Low-boiling aromatics are acutely toxic and high-boiling aromatics are causative agents for cancer.

Oil spills and sea discharges take their toll of sea birds and mammals. Oil, its components, and refined petroleum are toxic to fish, especially eggs and larvae. Reports of tainting of fish and shellfish are common in a spill's aftermath. Tar is fed upon by marine turtles and, along with plastic debris, is a cause of their dwindling numbers.

Oil washed ashore is devastating to intertidal life. If the animal is lucky enough to have a natural coating of mucus, like the sea anemone, the oil won't stick and it may survive. The same is true for some algae, but heavy spills usually wipe out the seaweed temporarily.

If the oil mixes with sand, it can persist for several years. On rock, it will weather to an asphalt coating and also last for years. Rocks covered with an asphalt coat rarely recolonize until the coating wears away.

Oil easily adheres to clay and silt. As it sinks, bottom sediments absorb it. Clams, abalone, starfish, sea urchin, lobster, and other bottom dwellers have been killed not long after spills. Small crustaceans also perish. When the bottom is heavily oiled, animal life is all but snuffed out. Recolonization is slow and, in the early stages, may only involve the hardier marine bristle worms. Full recovery can take five to ten years.

In the deep sea, petroleum makes its way into the sediments, carried down by sinking water, in fecal pellets and other detritus, and adhered to suspended matter. Hydrocarbon concentrations range from a few parts per million in sediments on abyssal plains to 70 ppm in coastal basins 600 to 1500 meters deep. (Shallow coastal water sediments may contain as much as 12,000 ppm). Aromatic hydrocarbons make up 60 percent of the total, both in mid-depth and deep water.

Little is known about the effects of oil on life at great depths, but combined with the cold, the low metabolic rates and specialized physiology of its animals, the potential for harm is severe.

INDUSTRIAL DISCHARGES

Coastal disposal of industrial waste has been steadily declining. Currently only fifteen major pipelines go directly into coastal waters and only three firms dump at sea (about 200,000 tons a year in 1985).

The coastal wastes cited by federal authorities as worrisome are oil-drilling fluids, inorganic acids and alkalies, pharmaceutical waste (aqueous suspensions from antibiotics production), and fish-processing waste.

PLASTICS

The persistence of plastics has created a whole new set of environmental troubles. Hardly a beach remains untouched by washed-up or blown-in plastic debris that refuses to rot.

Coastal states have had to mount new cleaning operations to keep their shores aesthetically pleasant for the tourist trade. The shorelines of New Jersey and Long Island are suffering from debris originating from sewage, landfill blow-off, and street and boat litter. To the south, and west into the Gulf of Mexico, all states report chronic beach litter and sporadic animal fatalities linked to plastic. The west coast of Texas possibly suffers the most. Material is carried there by inshore converging currents from the east and south, and by offshore winds. Thus, Padre Island is awash in the discards of Florida, Alabama, Louisiana, Mexico, the oil rigs, and all the gulf shipping.

At sea, plastics are creating problems for marine life and man. The small boatman experiences fouled props and jammed cooling intakes from drifting plastic. Sea wildlife dies from eating plastics or from entanglement. Lost nets and traps, now almost exclusively made of plastic cordage, continue to "ghost" fish long after they have been lost.

Little data exist on "ghost" fishing, the continual trapping of fish by open, submerged drift, gill and trawl nets, but occasional sightings have been made by scuba divers, who report the nets always contain fish. Lost traps also continue to fish. Each dying victim becomes the bait for the next. Derelict traps have been recovered full. Yearly trap losses by pot fishermen are sizable. In 1978 alone, 550,000 lobster traps were lost off New England. Biologists estimate those traps destroyed 1.5 million pounds of lobster that year.

The demise of several species of turtles has been partly blamed on plastic. Turtles try to eat it because it looks like jellyfish, a turtle staple. Declines in the numbers of northern fur seal in the Pribilof Islands are thought to be directly the result of net entanglement. Cape fur seals, California sea lions, elephant seals, harbor seals, monk seals, and Stellar sea lions have all been found ensnarled in nets and plastic strapping bands. Six-pack rings have proved fatal to thousands of seabirds.

The world's maritime fleets dump millions of tons of trash at sea each year. The federal government estimates that 690,000 plastic containers are tossed overboard every day. In the 1970s, commercial fishery fleets dumped 26,000 tons of plastic packing and lost 150,000 tons of plastic fishing gear yearly.

An international agreement to ban the disposal of plastics at sea (MARPOL-Annex 5) has been proposed, but it has not been ratified by the United States.

NUCLEAR WASTE

From 1946 to 1970, the United States dumped low-level radioactive waste offshore in steel drums at four sites in the Atlantic and two in the Pacific. Some radioactive waste has also leaked from defense sites, notably Hanford in Washington State, and washed downstream.

An indefinite moratorium now exists on low-level waste disposal at sea.

FUTURE OCEAN DISPOSAL

As land alternatives become harder to come by, pressure will mount to dump in the open sea. Present legislation (Marine Protection, Research, and Sanctuaries Act) has a legal hole in it that centers on the phrase "unreasonable degradation." The EPA has lost a court decision about the abolishment of ocean dumping. The judge held that EPA's "conclusive presumption that many materials which fail ocean environmental impact criteria will unreasonably degrade the environment was arbitrary and capricious." This has opened the door for further demands for ocean dumping permits. Philadelphia, Boston, Washington, D.C., Seattle, and San Francisco are all interested.

As many sewage plants convert to secondary treatment, more sludge will be created. Even if incinerated, the question remains of how to dispose of the ash. Municipal solid waste incineration is also increasing; it is another source of metal-laden ash. Conversion from oil to coal is producing great quantities of fly ash, flue-gas desulfurization sludge, and bottom ash (25–40% of all coal burned is ash). Companies are requesting permits to dump these at sea.

Unconsolidated ash can be very fine and likely to smother sea-bottom life. Coal ash can be converted into solid blocks. Experimental rubble reefs made of ash block colonize quickly and attract "wreck" fish and lobster.

Scientists like Willard Bascom argue that "inorganic metals in the sea are not a hazard to sea animals or to persons who eat them." He contends that not only is ash-dumping safe but the risk of dumping sewage sludge and dredge spoil, frequently cited as sources of heavy metals, is very small. Laboratory tests, he feels, using metals in forms infrequently found in nature and in concentrations much higher than those found in contaminated areas, are misleading.

He notes that animals around California treatment-plant outfalls exposed to sediments that contain mercury (less than 1 ppm) do not accumulate it. The mercury is in inorganic form and, apparently, is not being transformed into methylmercury at a significant rate. He also points out that sea animals have a protective mechanism against toxic metals. Their tissues contain a protein called metallothionin that mediates metal absorption and retention.

Deep ocean disposal of organic matter is not considered wise by scientists. They point out that the rate of decay at great depths is too slow and dumped material might accumulate and create some unanticipated mischief at a later date.

In the early 1970s, the "Alvin," a Woods Hole Oceanographic Institute research submersible, foundered and sank in 4500 feet of water. Retrieved ten months later, sandwiches in a water-soaked lunch box were remarkably well preserved. As has since been found, decay rates at great depths are 10 to 100 times slower than controls at the same temperatures at sea level.

In 1973, Arthur D. Little, a Boston-based consulting firm, prepared a study for the New England Regional Commission to find out the consequences of dumping compressed bales of municipal solid waste into the deep sea. Although part of the waste decays, much does not. Plastics eventually surface. The study could not predict the pollution effects on abyssal life nor the rate at which abyssal life, including bacteria, would attack the waste. It concluded that dumping solid waste into the deep sea was not a good idea.

As Gilbert Rowe (1982) has pointed out, the bottom of the deep sea is a fragile ecosystem, highly specialized, a seasonless, frigid, pitch-black, high-pressure, sparse environment, unaccustomed to insult.

The EPA has permitted California to continue to pump primary waste to sea. California's argument is that the high "assimilative capacity" of the open ocean supplies an "unlimited" amount of oxygen for waste decay, and the meager reduction in nitrogen and phosphorus during secondary treatment belies any benefit from the additional expense it incurs.

As for deep ocean sludge dumping, John D. Isaacs of Scripps Institute has pointed out that "six million tons of anchovies off the California coast produce as much fecal matter as 90 million people, ten times the population of Los Angeles." Critics retort that anchovies don't pass unbiodegradable toxic organic chemicals, toxic metals, or human pathogens.

Pressure is also mounting for the resumption of dumping of low-level radioactive waste—everything from contaminated clothing to soil. A panel of experts appointed by the Organization for Economic Cooperation and Development has concluded that marine dumping of low-level waste in the past has done no significant damage to marine life, has resulted in less human exposure than land-based disposal, and dumping at sea at higher rates than in the past would not have any adverse consequences.

The quandary about what to do with high-level nuclear waste has also led to the sea. Plutonium manufacture and commercial reactors have produced over 20,000 metric tons of hot waste that we must store for tens of thousands of years. No one knows how to do that with absolute safety. Where you put it must be geologically stable and impervious to any natural means of slow or sudden resurfacing. One notion has been to implant it beneath the sea in subduction zones, those borders where the expanding sea floor buckles under the more massive, adjacent continental plate. Another idea has been to bury it in the ocean's abyssal red clay, away from plate boundaries, where the slow rain of debris will increase its thickness a few millimeters every thousand years and continental drift will eventually carry it to the nearest plate boundary, then downward.

Both proposals have been criticized. Plate boundaries are unstable, given to earthquakes and volcanism. Deep bed water circulation might surface these radionuclides. Ocean circulation is slow. It takes a thousand years for mid-ocean water to surface, but, given the long lives of these hot wastes, that is altogether too soon.

SIX

THE MANAGEMENT MUDDLE

Laws grow from necessity. As air, water, and soil pollution worsened after World War II, federal statutes proliferated to deal with it. Between 1962 and 1977, more than thirty-five major pieces of environmental legislation were passed. Since then more have been added, and others revised, amended, and reauthorized.

Most of these laws have been designed so that the states can take on the responsibilities of oversight. This, in turn, has required that the states create authorities, programs, departments, and regulations.

In principle, this makes good sense. Problems of local origin are best solved by local authorities. Furthermore, the federal government is constitutionally limited in the extent to which it can directly intervene in state affairs. In practice, the response of the states has been uneven and, as federal funding has dwindled, the fates of a number of federally derived state programs are in limbo.

Laws have been enacted covering air, water, toxic waste, water supply, solid waste, and coastal development. Many of these regulations overlap on common problems. Consider sludge dumping, for example. If a municipality decides either to incinerate or landfill sludge formerly dumped at sea, five federal statutes apply: the Clean Air Act (CAA), the Federal Water Pollution Act, also known as the Clean Water Act (CWA), the Marine Protection, Research, and Sanctuaries Act (MPRSA), the Resource Conservation and Recovery Act (RCRA), and the Safe Drinking Water Act (SDWA). And, if siting an incinerator is an issue, don't overlook the NIMBY syndrome. In the past few years, the public has thwarted a number of municipal projects, often agreeing on their desirability but refusing to allow them to be located in their neighborhood. The acronym NIMBY stands for "not-in-my-backyard" and has become a source of environmental gridlock.

Currently, marine water quality is subject to twenty-one federal programs, under eight major statutes administered by eleven different federal agencies. Responsibility under these laws is divided by territory, by function, and by administration. As one past report summarized a sludge disposal issue:

> It was impossible to implement all five statutes simultaneously and, as a result, the implementation of each shifted the burden of receiving society's waste products to the medium that was least regulated at that moment. An industry or municipality faced with the problem of what to do with its wastes may find the CAA effectively prohibits incineration; the CWA similarly limits disposal at sea through a pipe or in internal waters by any means; the MPRSA prohibits disposal at sea via barging; and the RCRA and SDWA effectively prohibit land disposal or deep-well injection.

Standards conflict; for example, ocean quality criteria are more stringent than those for estuaries. Rules and jurisdictions overlap, as the sludge-disposal problem illustrates. Where the mechanics of the law are in place, monitoring and implementation have fallen short of intended goals. Municipal effluent still pollutes our major estuaries and waterways, and industrial dischargers still spew out dangerous materials. The goal of making all waters safe for recreation and wildlife is still far away.

MARINE BOUNDARIES

Marine boundaries determine the geographic limits of maritime law. A government or governments can implement the law no farther than the territorial limits of its authority.

The states have authority over waters within their boundaries and, more recently (1953), over their coastal waters, the territorial seas. The territorial sea extends from a low-water baseline outward three nautical miles.

Beyond the territorial sea lies, legally, the high seas. The high seas are open to any nation to traverse, to explore the seabed, and to fish. There are exceptions. A nation may extend its jurisdiction farther out for special purposes—the control of smuggling, defense, enviornmental protection, public health, and fisheries management, for example. In the past, this contiguous zone, as it is called, was limited to twelve miles off-shore.

Exceptions have proliferated. The 1958 Convention on the Continental Shelf gave each coastal nation control over its seabed to a depth of two hundred meters (about six hundred feet). Then, in 1976, the United States passed the Fishery Conservation and Management Act (the Magnuson Act), which extended control of fishing rights to the waters two hundred miles offshore.

INTERNATIONAL AGREEMENTS

The United States has not been a leader in promoting inter-national environmental agreements. However, it has entered into a few significant pacts.

Waste disposal on the high seas has been addressed by the 1972 London dumping convention, to which the United States is a signatory. It prohibits the dumping of organohalogens, mercury and its compounds, cadmium and its compounds, oils, radioactive materials, and agents of biological and chemical warfare. Besides this "black list," there is a "gray list" of compounds that can only be dumped by special permit. What is not on either the black or gray list still requires a general permit from either the flag nation or the nation in which loading takes place. The London convention is aimed principally at wastes originating on shore.

Dumping oil or refuse at sea as a part of normal ship's routine has been addressed by the "Protocol of 1978 Relating to the International Convention for the Prevention of Pollution from Ships, 1973," MARPOL 73/78. The United States has not ratified the optional annexes that cover dumping of hazardous packaged materials, ship's sewage, and garbage.

European and other nations have a series of agreements of a regional nature—the Oslo, Paris, Barcelona, and Helsinki conventions and the Bonn Agreement—that regulate land-based discharges, ship dumping in adjacent waters, and oil pollution. The 1979 Kuwait Convention focused specifically on oil pollution from tankers, refineries, and petrochemical industries.

FEDERAL LEGISLATION

Three major statutes now control the fate of our estuaries and coastal seas: the Clean Water Act, the Marine Protection,

Research, and Sanctuaries Act (MPRSA), and the Coastal Zone Management Act (CZMA). There are a number of others. These are briefly described in Appendix A.

The Clean Water Act

The Federal Water Pollution Control Act, better known as the Clean Water Act (CWA), has jurisdiction over U.S. waters, including the coastal sea to the three-mile limit. It was passed in 1972 and has undergone considerable modification and expansion since then. Congress established its primary purpose as the "restoration and maintenance of the chemical, physical, and biological integrity of the national waters." Its original goal was the elimination of the discharge of pollutants into navigable waterways by 1985 and an interim goal of providing water with quality sufficient for the propagation of fish, shellfish, and wildlife and for recreation by mid-1983.

It established as a national policy the elimination of discharges of "toxic pollutants in toxic amounts," federal financial assistance for construction of publicly operated waste treatment plants, areawide waste-treatment management planning, the development of technology to "eliminate the discharge of pollutants into navigable waters, waters of the contiguous zone, and the oceans and programs for the control of nonpoint source pollution." Responsibility for implementing and adminstering CWA programs is delegated to the states once they have established programs to implement federal law.

Federal grant programs have been established to help municipalities create pollution-control programs to regulate both industrial and municipal discharges. The regulations require that all industrial and municipal direct discharges into navigable waters obtain a National Pollution Discharge Elimination System (NPDES) permit. The discharge must comply with effluent limitations, not violate water-quality standards, and, for direct marine discharges, meet ocean quality criteria.

With a few exceptions, municipal plants must upgrade to secondary treatment. Industries must meet quantity-discharge limits based on EPA guidelines. For industries discharging into public treatment plants (indirect dischargers), municipalities must develop programs to assure that these wastes receive pretreatment. Standards prohibit effluent that can cause a fire or explosion or interfere with plant operation. Categorical pretreatment standards are now in place for major types of industries.

Initially, EPA concentrated on conventional pollutants: suspended solids, BOD, oil, grease, coliform bacteria, and the like, but a suit brought by the National Resources Defense Council insisted that EPA pay attention to "non-conventional" and toxic pollutants. Under what is known as the Flannery Decree, EPA developed a list of these pollutants and established rules for dealing with them.

Since 1977, EPA has set treatment standards for most primary industries. Industrial dischargers were to have achieved a basic level by July 1977 and to have moved on to the use of more sophisticated treatment technology by July 1984. A few have done so; "Organic Chemicals, Plastics and Synthetic Fibers" manufacturers have achieved 98 percent reductions, but other groups such as "Metal Finishing and Plating" have lagged far behind. Indirect dischargers, thus far, have also been slow to comply.

The Clean Water Act also covers the disposal of dredge spoil. Historically, the Army Corps of Engineers (COE) has had the authority to regulate activities in rivers and coastal waters that affect navigation. COE evaluates permit applications using guidelines developed with EPA, the Fish and Wildlife Service, the National Marine Fisheries Service, and the states. EPA can veto any proposed disposal sites for dredged material.

All NPDES-permitted discharges from point sources into territorial seas, the contiguous zone, or open ocean must not "unreasonably degrade the marine environment," according to the Clean Water Act. Unfortunately, this rule does not apply to discharges into estuaries and certain coastal waters. The definition of "unreasonable degradation" is:

• Significant adverse changes in ecosystem diversity, productivity, and stability of the biological community within and surrounding the discharge area;

• a threat to human health through direct exposure or through consumption of exposed aquatic organisms; or

• loss of aesthetic, recreational, scientific, or economic values that is unreasonable in relation to the benefit derived from the discharge.

Secondary treatment of treatment-plant waste can be waived, but the municipality must implement pretreatment by indirect industrial dischargers and comply with existing water-quality standards. The waiver was intended primarily for West Coast municipalities that discharge into deep, cold waters.

Because so many statutes and federal programs impinge on estuaries and coastal waters, Congress provided EPA with the role of coordinator. Under the National Estuary Program, EPA conducts estuaries-management conferences for data-gathering, monitoring, and setting priorities. These conferences also address other forms of watershed pollution and state management.

Each amendment and reauthorization of the Clean Water Act has contained a long list of appropriations and decrees.

The Water Quality Act of 1987, for example, directed that the New York sludge dump site be moved to a deep-water site, took aim at specific estuary improvement (Chesapeake Bay, Boston Harbor), provided treatment-plant construction grants, and addressed toxic pollutants in sludge, "hot spots," and nonpoint sources. It also increased penalties for civil and criminal violations, and again extended compliance deadline for industries.

The major provisions in the CWA are summarized in Appendix B.

MPRSA—The Marine Protection, Research, and Sanctuaries Act

MPRSA, or the Ocean Dumping Act of 1972, as it is also known, regulates the transportation and dumping of waste offshore. It applies to dumping of material originating in the United States, from U.S. vessels in either the territorial sea or the contiguous zone and open ocean. It overlaps coverage with the Clean Water Act in territorial seas and preempts it. Thus, coastal and open-ocean dumping fall under MPRSA and dumping in estuaries is under the control of CWA.

MPRSA is the only U.S. pollution law that explicitly requires that the consequences of alternative disposal methods be considered, which, in effect, means that the problem cannot be simply transferred from one arena to another without at least some forethought.

The law governs the sea dumping of solid wastes, incinerator residues, sewage sludge, industrial wastes, dredge materials, low- and high-level radioactive wastes, and chemical and biological warfare agents. High-level radioactive waste and warfare agents are specifically prohibited from ocean dumping. Effluents discharged through outfalls, and oil or sewage discharged from vessels are not covered by MPRSA.

Four federal agencies are involved: the Environmental

84

Protection Agency (EPA), the Corps of Engineers (COE), the National Oceanic and Atmospheric Administration (NOAA), and the Coast Guard.

EPA designates the specific disposal sites, runs the permit system for using the sites (with the exception of dredging), and sets the criteria for the disposed material. Permits for dredging are controlled by the COE, although EPA has review authority.

Both EPA and NOAA are required to conduct research and monitoring on the effects of ocean dumping and to study alternative disposal methods.

The Coast Guard is charged with the surveillance of ocean dumping.

In a 1986 reauthorization of this law, an amendment directed EPA to cooperate with other government agenices in assessing the feasibility of regional management plans for waste disposal, the idea being to integrate all waste disposal into a single, comprehensive strategy.

MPRSA also gives the Secretary of Commerce the authority to set up marine sanctuaries as far seaward as the continental shelf.

The original goal of EPA in 1973 was to ban all ocean dumping and, indeed, industrial dumping has significantly decreased. Many major cities and municipalities phased out sludge dumping, but, on the whole, it has increased, mainly due to inputs from the New York metropolitan area.

Several catastrophic environmental incidents in the mid-1970s (a massive bottom kill in the New York Bight, for one) prompted Congress to mandate a ban on all "harmful" dumping by 1981.

In 1981, New York City brought suit against EPA to halt implementation of the ban. The Federal District Court in New York ruled dumping could not be banned without fully considering alternative environmental consequences and costs.

According to the court, EPA's presumption that materials that fail ocean environmental impact criteria will "unreasonably degrade" the environment was "arbitrary and capricious." Thus New York and other metropolitan municipalities continue to dump, although the site has been moved from the old 12-mile site to 106 miles offshore.

EPA did not appeal the 1981 court decision and appears to be shifting its policy from ocean protection to management. However, it has not permitted additional dumping even though a number of coastal cities are interested in doing so.

The Coastal Zone Management Act

In 1972, the Coastal Zone Management Act (CZMA) was passed. It addressed physical degradation on U.S. coasts from overdevelopment. The goals of CZMA are "to preserve, protect, develop, enhance, and restore, where possible, the coastal resources." To do this, the law was designed to "encourage and assist" state and local governments, develop interstate authorities, and supply the cooperation of federal agencies.

To encourage state participation, federal incentives included the promise that federal actions involving the coastal zone would be "consistent" with state programs and that the programs would be supported with federal funds. Grants can be awarded to coastal states for the process either of developing programs or implementing them.

To be eligible, the state must set up a management program that identifies the coastal boundaries and the land and water uses it will permit within the coastal zone, classify areas and set use priorities for areas of particular concern, identify and describe the authority, structure, and procedures it will use to operate the program, define beaches and public access plans, plan for energy facilities (power plants), and for coastal erosion.

The state must also provide a way of coordinating with local governments by notifying them when a state decision conflicts with a local ordinance and designating a specific state agency to carry out the program.

Although laudable for its goals, the Coastal Zone Management Act has had few tangible results. Rather than fostering resource conservation, it has drifted into the murky waters of procedure and management, producing little of substance.

State program development has been slow and patchy; six states have not participated at all, and it took ten states more than eight years to get the program through their own legislatures and win EPA approval.

Turf wars and power struggles at the state level have taken the teeth out of some programs. For example, the North Carolina General Assembly nearly killed its coastal program in 1977 over the issue of who designates "areas of environmental concern." They succeeded in substantially watering down the power of the Coastal Commission and, to date, development is rampant in environmentally sensitive areas, such as Bogue Banks, a long barrier island. The situation is far from bleak on many other North Carolina coastal issues, but pressure from developers is intense as it is in many coastal areas.

State and federal conflicts have arisen over the "consistency" provision. Federal projects and actions must be consistent with state program goals. Thus, Corps of Engineers dredging permits and federal permits for energy facility siting must meet these criteria. Offshore oil development has been pushed by the Interior Department in the face of disapproval of Alaska, California, Florida, and Massachusetts. Also, the Corps of Engineers have been at odds with many state authorities over spoil dumping.

Federal funds have dried up during the Reagan administration. The 1982 Reagan budget called for the elimination of all funding for CZMA, saying the program "has largely achieved its purpose." Congress thwarted the move, but federal money has been in short supply ever since. In 1987, $34 million was all that was available for twenty-nine states.

GOVERNMENT ON HOLD

At the time of Ronald Reagan's election in 1980, the EPA, which had been created in 1970 under the Nixon administration, was the most powerful regulatory agency in Washington. It had nearly 15,000 employees, seven regional offices, research laboratories in twenty states, and responsibility for the implementation of nine major laws.

Whether it was the enormity of the tasks it faced, the political tenor of the times, or an internal operating philosophy, EPA was slow to demand compliance with the law from state and local governments. National Pollution Discharge Elimination System (NPDES) permit violations by waste-treatment plants are an example. A study begun in 1979 by the General Accounting Office, and submitted to Congress in November 1980, found that 87 percent of 242 randomly selected treatment plants were violating their NPDES permit discharge limitations for at least one month of the year and 31 percent were in "serious" violation. By that time, the federal government had invested $25 billion in construction grants for plant upgrading, and EPA wanted an additional $36 billion. The GAO report highlighted the inefficiences of the plants built or upgraded with that money.

The Reagan administration's appointments of James Watt as Secretary of the Interior, Anne Gorsuch as administrator of EPA, and David Stockman as director of the Office of Management and Budget laid the groundwork for the reduction, elimination, or suspension of hundreds of regulations and the introduction of a "new mindset" in EPA management.

Three months after Ms. Gorsuch was confirmed (May 1980), she eliminated the EPA enforcement office, folding their personnel into program offices. Not that EPA's enforcement record had been anything but mediocre to that point. EPA simply had not been enforcement-oriented, wanting, as one observer put it, "to retain a nice-guy image."

The transition was chaotic. Word of a "hit" list got around and the agency began to lose personnel from both attrition and purges. Morale plummeted. Enforcement of existing laws ground to a halt. EPA regions had referred 313 cases for legal action in 1980, 59 in 1981. The word had gone out that "every case you refer to headquarters is a black mark against you."

The Gorsuch era ended with her resignation on March 9, 1983, in a controversy over the refusal of EPA to release records to Congress. The EPA staff had been reduced by 30 percent and their budget slashed.

The return of William Ruckelshaus (the first EPA administrator, in the Nixon years) failed to turn the tide. He could not restore funding cuts, leaving the agency overworked and understaffed.

TOWARD A NEW POLICY

Some observers think there are more than enough laws now on the books to clean up the country. The stumbling block is the reluctance to implement and enforce them.

Others are not quite so sure. The gist of present environmental law, they say, is reactionary—one of containment and treatment rather than prevention. Massive treatment may be the most practicable course to follow to meet immediate problems, but for the longer run, it will be too expensive and too burdensome on society. It is far easier and cheaper to neutralize, extract, and recycle, or eliminate waste at its source.

Human waste, as well as animal waste, unadulterated by toxins and heavy metals, can be recycled to the land. Fresh water can be better conserved if less is used to carry off waste. Industrial pretreatment, removing organic chemicals and heavy metals before they enter municipal waste streams, would allow more options for water recovery.

Clean rivers mean less costly and safer drinking water and more water available for other purposes. Clean estuaries open possibilities for aquaculture and expanded production of fish and shellfish and the recovery of waterfowl and other marine wildlife.

To accomplish this requires generating and sticking to a series of long-range precepts about how we will change our infrastructure. Just as the old septic systems became outmoded as population densities increased, so has our current waste technology become obsolete given the evolutionary direction of our society. Combined sewage overflows must go. Even the combination of municipal and industrial waste must be called into question in high-density areas. Coupled with crises in air quality and solid-waste control, an overhaul in total policy is quickly becoming a national priority.

SEVEN

THE AQUAPHILE'S AGENDA

Leading the fight for estuarine waters is a patchwork of private, nonprofit conservation groups. A few have large memberships, but, on average, most are small and operate with minuscule staffs and low budgets. In October 1987, leaders of sixty-five coastal activist groups gathered in Rhode Island for a five-day workshop hosted by Save the Bay and sponsored by the EPA and the Belton Fund. The upshot of the conference was a unified position on major environmental issues and the selection of those needing immediate attention.

The delegation chose five top coastal priorities:

- The nation must adopt a subsidy-busting program.
- The nation must adopt a source-reduction program.
- America needs to save its wetlands.
- The nation needs an aquafund program.
- The nation needs environmental law and order.

SUBSIDY-BUSTING

Too many federal programs encourage and subsidize development in environmentally foolhardy places such as barrier islands and flood plains. The major culprit is the National Flood Insurance Program, in which the taxpayer underwrites $163 billion in policies for flood insurance in unstable and repeatedly flooded locales that private insurers won't touch. The program is second only to Social Security in national liability commitments. It has led to far more new construction in high-risk regions than would have otherwise occurred if prospective owners had to bear the insurance burden by themselves.

SOURCE REDUCTION

What we do with waste must be accompanied by action plans to reduce and recycle as much of it as we can. "Out of sight, out of mind" solutions solve nothing.

The better bet is not to create waste in the first place. What cannot be reduced can be partly recycled.

Industrial waste should be recycled or destroyed as close to its point of origin as possible and not mixed with municipal waste where it becomes basically untreatable and can interfere with biological processes.

SAVING OUR WETLANDS

The rapid disappearance of our wetlands and the degradation of what remains must be halted. Nationwide, 54 percent of our wetlands have been filled in and over half of the remainder are imperiled by physical and chemical destruction. Concurrently, shore-based buffer zones are needed to protect wetlands and adjacent waters from land runoff and other effects of high-density development.

AQUAFUND PROGRAM

Toxic "hot spots" and contaminated sediments in harbors and waterways must be cleaned up. Some areas are polluted heavily enough to be eligible as Superfund sites—an EPA program for emergency action on the most seriously polluted sites—if the law covered waterways. A fund must be established to remove these sediments and rejuvenate afflicted areas.

ENVIRONMENTAL LAW AND ORDER

Current legislation and regulations are a bewildering potpourri of fragmented and overlapping responsibilities that are neither fully implemented nor enforced. For the present, federal, state, and local laws must be firmly enforced. For the longer term, a strong, coordinated, and self-consistent environmental policy must be invoked, implemented, and enforced.

The three working groups' recommendations go into more detail:

Controlling Water Pollution

Point-Source Pollution

1. *Discharge permits should be issued only after a "need to discharge" has been demonstrated.*

The National Pollution Discharge Elimination System (NPDES), through permitting, controls point-source discharges. The agency should issue permits only if no other options are available and, if granted, the grantee should assure the agency that the best possible pretreatment system is in place.

2. *The nation must establish a source-reduction program.*

Initiate incentives to reduce toxic waste by process changes and good housekeeping. Fund it from penalties collected by Clean Water Act enforcement.

3. *Comprehensive standards for water, sediment, and sludge should be promulgated to protect our waters.*

Under NPDES now, standards exist for only a few dozen pollutants. The EPA plans to regulate only 126 "priority" pollutants. Harmful unregulated pollutants should be brought under Section 307 of the Clean Water Act. So should standards for sediment and sludge; none exist now.

4. *Permits should be based on the above standards with no consideration of mixing zones.*

Water quality standards are now based on the effects of dispersed pollutants and do not take into consideration local damage by concentrated effluent.

5. *Permits should be designed to address all effects of pollutants on a water body.*

Currently, combined waste-load "allocation" covers only multiple-point sources and ignores the additional burdens of air fallout, runoff, and sediments.

6. *Comprehensive monitoring programs should be established to evaluate the effectiveness of discharge limitations.*

Waters receiving discharges should be monitored by NOAA's Status and Trends program and EPA's National Estuaries program. Note of any ill effects should trigger immediate remedial action irrespective of whether established standards are violated or not.

7. *Permit fees should reflect the true cost of disposal.*

Fees should be levied based on the toxicity of the effluent, which should create an incentive for reduction. Fee funds should be applied to clean up existing problems.

8. *Adequate funds from local, state, and federal governments should be raised for the construction of publicly owned sewer systems.*

Public law dictates that secondary sewage treatment is mandatory. Currently, many treatment plants go no farther than primary treatment. For some sensitive receiving waters, tertiary treatment may be necessary.

Non-point Source Pollution

Polluted runoff, urban and agricultural, must be stemmed by "Best Management Practices" (BMPs) such as conservation tillage, animal-waste control laws, street cleaning, etc.

The Federal Construction grants program should be renewed. More Clean Water Act funding is needed. State funding must be encouraged.

9. *Comprehensive strategies must be developed to elim-inate adverse effects from sewage runoff.*

During rains, combined sewage overflows (CSOs) bypass the treatment plant and dump raw waste. The cost of replacing these older systems is high, but some way of controlling these discharges must be found.

10. *Require implementation of "Best Management Prac-tices" in coastal areas.*

To control agricultural runoff, target the offending areas. Man-date and monitor USDA BMPs and be sure BMPs maximize benefits to water quality. Tie farm subsidy support to com-pliance.

11. *Source-reduction campaigns to reduce nonpoint pol-lution should be required.*

Agro-urban runoff can be reduced by source reduction. Curb the use of fertilizers and pesticides and provide "how-to" re-placement assistance.

12. *Redirect the conservation title of the 1985 farm bill to emphasize water quality and habitat protection in coastal areas.*

The Conservation Reserve program, designed to retire 45 million acres of cropland, is not taking hold in coastal areas. The Department of Agriculture should extend current con-tracts from ten to fifteen years and provide incentives to plant trees. CRP should expand to encompasss "water quality" lands as well as "highly-erodable" acreage.

13. *Mandatory recycling programs must be adopted in all coastal areas.*

Solid waste is finding its way into coastal waters. Adopt waste-reduction and recycling programs aggressively.

14. *More extensive education must be provided on coastal processes and habitat needs.*

Coastal ecosystems should be part of school science courses. More information should be made available to adults.

Air Toxics

15. *Clean Air Act reauthorization should include controls on sources of airborne toxic pollutants.*

The U.S. Senate is considering a reauthorization bill that will require the EPA to control toxic air pollution by:

- establishing Best Available Control Technology for 224 toxins;
- providing regulations for leak prevention and detection of toxic substances;
- issuing regulations to reduce pollution from small sources such as dry cleaning establishments;
- listing certain pollutants as "extremely hazardous" and requiring EPA to set very stringent standards.

16. *As the Federal Insecticide, Fungicide, and Rodenticide Act is considered for reauthorization, all pesticides which have been identified in fish and wildlife tissues or*

have the potential for atmospheric dispersion should be restricted in the licensing process and allowed to be used only with state controls in application and drift.

17. In all areas of waste management, source reduction must be made the top priority.

18. State nonpoint pollution-source assessment plans should include atmospheric sources.

19. The EPA and NOAA should be required to conduct national assessments to document the sources and effects of air pollution on coastal waters.

(Recommendations 18 and 19 have become all that more timely because of the recent disclosure by the Environmental Defense Fund study showing acid rain to be a major nitrogen contributor to eastern estuaries. See comments under item 21.)

20. The EPA and Congress must integrate pollution control strategies regulating all sources.

As bills come up for reauthorization, regulations should be redesigned to control toxins from all sources. The Toxic Substances Control Act (TOSCA) should be expanded to put the burden of proof on manufacturers to show environmental acceptability of compounds before they are licensed for production.

21. Public education is needed in order to effectively control air pollution and its impacts on our waters.

Since the conference, the Environmental Defense Fund has released a report on the overenrichment of coastal waters by acid rain. Emanations of sulfur dioxide and nitrogen oxides are converted in the atmosphere into sulfuric and nitric acids that are carried earthward by rainfall. The major sources are coal-burning public utility plants and automobiles. The surprise has been the quantities involved. Acid rain is second only to fertilizer runoff as a nitrogen source in Chesapeake Bay. It contributes one-fourth of all nitrogen generated by human activity, exceeding the output of sewage outfalls and animal waste.

Sediment Contamination

22. *A new "Aquafund" program for in-place pollutants should be established.*

Presently, polluted sediments in our waters are not recognized by the Superfund program. Create a national program to identify, rate, and clean up contaminated sediments.

23. *The Army Corps of Engineers and the EPA should revise current dredging and dredge spoils disposal criteria.*

Prohibit water and wetland disposal of contaminated dredge spoil.

Offshore Sources

24. *Ocean dumping of wastes is not a preferred management option and should be phased out.*

25. *EPA should abandon its ocean incineration program and redirect resources into existing and emerging innovative technology.*

EPA curtailed its ocean incineration program as of February 1, 1988.

26. *Congress should enact comprehensive cleanup standards for oil and chemical spills from vessels and offshore facilities.*

The United States should ratify the 1984 Protocol to the International Civil Liability and Fund Conventions. These establish a liability and compensation code for oil-spill victims.

27. *The United States should ratify and implement all optional annexes of MARPOL.*

The International Convention for the Prevention of Pollution from Ships (MARPOL) regulates five categories of ship pollution: oil, bulk chemicals, hazardous substances in packages, sewage, and garbage.

Land Planning: Preventing Further Degradation

Coastal Development Pressures

1. *Eliminate government expenditures and subsidies that promote environmental harm in coastal areas.*

End economic incentives to dredge, fill, and develop coastal lands. The National Flood Insurance program should not encourage building or rebuilding in storm- and flood-prone areas.

End agricultural tax credits, deductions, and subsidies for draining wetland and irrigation projects. Stop federal projects—water diversion, channeling, drainage—that harm coastal habitats. Redirect the USDA's Small Watershed Program to benefit water quality and reduce soil erosion.

2. Development must be limited to areas that provide adequate services. No new growth should occur in undeveloped and environmentally sensitive coastal areas.

The Coastal Zone Management Act of 1972 (CZMA) was meant to protect sensitive areas. State and local land-use plans are not stopping degradation and overdevelopment. Either strengthen CZMA or create stronger regulations.

3. Provisions of federal law must be strictly enforced to maintain existing water quality classifications.

4. The Coastal Barrier Resources Act (CBRA) must be expanded to include all nondeveloped coastal areas including those of the Pacific coast and the Great Lakes.

This 1982 law disqualifies federal development subsidies, including flood insurance, on undeveloped barrier lands. Remaining undeveloped coastal areas should be included. Presently, the Pacific coast and the Great Lakes are not.

5. Mandatory buffer requirements must be established to adequately protect coastal habitat.

Wetlands and shoreland edges should be left undeveloped (no bulkheads, levees, or seawalls), and a 1000-foot buffer zone established.

6. Greatly expand government programs for open space preservation and increase opportunities for public access to coastal areas.

7. In sensitive coastal areas all deeds must include an environmental assessment of the property, highlighting the land's unique values and any restrictions on its use.

A potential buyer should be aware of land-use regulations affecting his proposed purchase.

Physical Alterations of Water Bodies and Watersheds

8. A comprehensive national wetland habitat protection act is needed.

9. Comprehensive state wetland protection programs must be established.

Few states now have them.

10. Federal and state review procedures must be established for oversight of local decisions impacting bays, sounds, and the Great Lakes.

The CZMA, the National Estuaries program, or the Clean Water Act can not now overcome parochial decisions or neglect at the local level.

11. Comprehensive strategies for protection and restoration of beach and shoreline areas must be developed.

12. *A National Watershed Act must be developed.*

The act should require an environmental-impact statement for any hydrological modification within a watershed or estuary; fund and create fish- and wildlife-restoration programs; protect existing free-flowing rivers from further channeling and navigation projects; initiate a national dam-removal-evaluation project.

13. *All streams, rivers, and lakes must be guaranteed flow rates that protect fish and wildlife habitat, water quality, and recreation.*

14. *Section 208 of the Clean Water Act must be implemented.*

Section 208 requires that all watershed sources of nonpoint pollution be taken into account when formulating regulations for the protection of a water body. This has rarely been done.

15. *Uniform national instream-flow measuremnt standards for fish and wildlife must be established.*

Withdrawing and diverting water can significantly alter estuary salinity and harm wildlife.

16. *Tighter coordination and accountability for downstream impacts must be forged among watershed agencies.*

Although the conference participants covered a number of waterfront issues of immediate concern, larger, more pervasive dilemmas remain.

This nation has no coherent, consistent, and enforced overall plan that addresses either our current or our future waste problems. Nor do we have a firm, deep-felt environmental conscience to guide us.

The improvement of our waters, both fresh and marine, will require more stringent laws and enforcement. It will impose restrictions we are not accustomed to, and costs we've never had to bear before. It will hurt. But land, air, and water are no longer infinitely abundant.

What we once regarded as free has grown scarce and abused. For our own well-being, these new restrictions can come none too soon.

EIGHT

IT'S YOUR MOVE

For every hundred people you find who lament the destruction of coastline waters, you will find only one who decides to do something about it. Lena Ritter of Stumpy Point, North Carolina, is such a one. Seven generations of her ancestors were shellfishermen. She, her husband, and their friends made their living from the waters of Pamlico Sound. In December, 1982, a newspaper article disclosed plans for development of nearby Permuda Island—383 condos, 4 tennis courts, 2 swimming pools, and a marina. All this right next to their leased shellfish bed!

At that time, Ms. Ritter couldn't name a county commissioner, had never attended a public meeting or made a speech, and knew little about the local ordinances, state agencies, or pertinent federal laws. But the proposed development spelled trouble for the adjacent waters and she knew her livelihood was threatened. She began a crash course in civic participation.

Because the coastal zone resources are so limited and the

demands of people living on the coast so diverse, state and federal laws include provisions for citizen participation.

The majority of coastal states now have a coastal management commission, a staff organization to support that commission, and an environmental or conservation agency concerned with other coastal problems such as water quality. Your first order of business is to find out who does what and how that system works.

Lena Ritter did. She and the group she formed successfully blocked the development (and the state eventually purchased the island!) and are now involved with other threats to their community.

LEARNING THE ROPES

If your concern encompasses a new, sudden, and timely issue, chances are others nearby are as concerned as you, some enough so to form organized resistance. If your worry is an old and nagging one (and most water quality problems fall into this category), you may be able to find an existing organization with which you can work.

Begin your search by looking for a local group, either independent or a chapter of a larger one, whose aims fit yours and whose specific goals match your interests. Find out what they know, what they have done in the past, what they are doing now, and what you can learn by working with them.

Small groups, by their nature, have few resources and, to be effective, must doggedly pursue narrow goals. Many are single-issue-oriented or geographically limited. Don't expect them to embrace a new challenge unless it lies in their turf. Don't expect to dump the issue on their doorstep without bringing along the resources to deal with it.

Larger organizations with broad interests, also perennially short on funds and long on agenda, have the opposite problem. Often, they cannot get embroiled in specific, local issues simply because there are so many of them. Their chapters may be able to get involved, offer guidance, or put you in touch with more likely prospects.

You can, of course, form a new organization either with a specific goal in mind or with broader implications. It's a little late for the latter and usually takes someone with past experience and the ability to engender the wide support needed to sustain it. Money and time become critical early on.

Putting together a group to face a single, well-defined, local issue is easier, more direct, and often, more fruitful. The obvious choice of recruits is people like yourself, those adversely affected by the issue.

Search for those who will actively help. You need two types: doers (everything from stuffing envelopes to writing articles) and advisors—legal and scientific professionals who can lay out the choices and show you how to implement them. From this group, a small central cadre must emerge that defines group action: who, how, when, and what the group will do and, as important, what it will not do.

Day-to-day concerns will rapidly go beyond your conservation goals. You will face some of the same problems that trouble a small business. You might do well to keep the Pareto Principle in mind. Problems come in two distinct forms, the vital few and the trivial many. Spend 80 percent of your effort on the former and no more than 20 percent on the latter. If you cannot distinguish between the two, be a follower, not a leader.

GET INFORMED

The federal laws most pertinent to coastal cleanliness are based on the Clean Water Act and CZMA. Write your regional EPA office for information. Because these laws have been on the books for so long and so much has been written about them, be reasonably specific about what you want. EPA has numerous publications on individual programs (eg., the National Estuary Program) that can be of help. For the Clean Water Act, the best single source is "The Clean Water Act of 1987" published by the Water Pollution Control Federation (see address in "Further Reading.").

The blue pages of your telephone directory or the "Government" section of the white pages will list state and federal agencies. In some states, there is more than one agency involved; one implements the Clean Water Act, the other, CZMA. Call them and ask for copies of state law applying to implementation of CWA and CZMA. You may find yourself switched from one extension to another before you find someone who understands what it is you are after. Keep cool and be pleasantly persistent.

When you get help, get the name, address, and phone number of the person who helped you. If you don't get what you want, you have a starting point from which to continue and, if you do get what you want, a thank-you note never hurts.

Agency rules and regulations are based on interpretations of state law which, in turn, are based on interpretations of federal law. Read as much as you can about your issue. Talk to agency people about what rules mean. Pose the same questions to more than one staff member. You may get different responses either because one or the other interprets them differently or one can explain them better.

Check you own interpretation of regulations with them. If you do not agree, is it because you do not understand the rules, or is it a matter of viewpoint?

How the legislators viewed their work is often on the record somewhere. The preamble to federal regulations, published in the *Federal Register*, states the intent of the law.

For state legislation, public hearing transcripts or minutes of commission meetings can clarify their original intent.

Government operates through two distinct channels: elected or appointed officals, and permanent staff. Both are important elements in getting your concerns met. As with all human contact, don't expect everyone in government to be "user friendly" without your expending time and effort to develop relationships.

If you can arrange to meet informally with agency people you have reached by telephone or met at public hearings, so much the better. Don't expect much interplay on a one-on-one basis at public hearings. It is seldom a format for more than formal statements, but will give you an opportunity to meet some agency personnel.

Elected officals—local, state, and federal—can be key elements in your efforts to get new legislation or get existing legislation enforced.

Do your homework. Who are your elected representatives and what is their voting record on environmental issues? The League of Women Voters can help you on both counts.

Congressmen maintain offices in their districts. They and their staffs welcome visits from citizens. Get to know them.

THE WAY THINGS WORK

Regulations protecting the environment have their origins in federal or state legislation. These laws are implemented by state or federal agencies who develop and adopt rules and requirements for meeting those legislative goals.

To develop the rules and regulations, agencies must interpret the law. Generally, the law is broad-based and does not spell out every conceivable situation to which it might be applied. The agency cannot foresee all the contingencies, either, and must continually adapt to new situations as they arise.

Agencies must decide what to do where the law is vague or ambiguous. Public advocacy can be crucial in swaying those decisions. If you feel an agency or a commission has misinterpreted the law, you can file a lawsuit and let the courts decide the law's interpretation. If you do not agree with the court's interpretation, you can appeal or try to get Congress or the state legislature to change the law.

States differ in the ways in which they administer environmental matters. Some have divisions consisting of an appointed commissioner and a staff whereas others have a department headed by an individual reporting directly to the governor. Some states have several divisions, each handling a segment of coastal issues, either in a single department or split between several departments.

Whenever a commission or department proposes a new regulation, federal law provides for public participation, generally through hearings. Public notice of the hearings and the text of the proposed regulations are published in the state's register (for state agencies) or in the *Federal Register* (for federal agencies). Notices of hearings may appear as legal ads in newspapers or be mailed to people who have previously requested to be notified of regulation changes for a specific agency. The notice will tell you how to get more information on the proposal.

You may submit written comments before the hearing (a worthwhile practice), sign up to speak at the hearing, and submit comments after the hearing.

Try to submit your views in writing whether you speak or not. State agencies are required to prepare a report of the hearing and make it available to the public and federal officials who review state programs.

In most states any citizen can petition an agency to adopt or change a rule or regulation. If you choose to do so, find out the procedure from the agency. In North Carolina, for example, the agency must respond to the petition within 120 days. They may (1) initiate the rule, (2) deny it, or (3) defer it to a later date (if you concur) to have time to study it.

Permits

Under both the CWA and CZMA, rules and regulations require permits for operations that may affect the environment. When an agency grants a permit, it implies that the agency has found the activity environmentally sound.

Permits can be general or specific. A general permit is a vehicle for rapidly approving ordinary projects, easily carried out within regulations with little or no environmental threat. General permits may not even require a formal request and do not require a review process or public notice.

If, however, someone wants to perform an activity that impinges on an environmentally sensitive area, he or she must obtain a specific permit to do so.

Public participation varies with the type of permit. Some require public notice of the application and a public hearing, if interest warrants. Others provide for written comment but no public hearing. Still others call for neither public notice nor comments.

If you think a permit has been wrongfully issued, you can challenge it by filing an appeal. Depending on the kind of permit or the circumstances, the appeal will be made either to the appropriate commissioner or to a court of law. You will need an attorney and, probably, expert witnesses. Occasionally the process is called a public hearing, but in reality, it is a contested case hearing.

Enforcement

Federal and state agencies have the authority to impose fines or other civil penalties on those who violate laws and regulations. If the agencies can show willful violation, they can impose criminal penalties. Most environmental penalty clauses are so written that each day of noncompliance is a new violation, so penalties can accumulate rapidly.

An agency, through the courts, can also seek an injunction to stop violations and repair the damage done. The violator and the agency, through a court order, may enter into a consent agreement that sets a timetable to correct the problem.

Enforcement requires monitoring. Some monitoring is voluntary, done by the potential violator (e.g., NPDES permits); other monitoring is the responsibility of the government. More and more, environmental groups have turned to direct or indirect monitoring to prompt action. Either they watch environmentally sensitive areas for infractions or look at the records of those discharging waste.

All point sources of pollution require an NPDES (National Pollution Discharge Elimination System) permit. The permit spells out the quantities and concentrations of specified pollutants that can be legally discharged. The permit holder must submit monthly reports about the quantities and characteristics of their effluent. The law, CWA section 101(e), mandates that these reports be made available to the public. In fact, the law encourages "public participation in the development, revision, and enforcement of any regulation, standard, effluent limitation, plan or program . . . under this act."

These records are reasonably accurate. To falsify them is a criminal offense. They are available, either from the EPA regional office, the appropriate state agency (if the state has been granted authority to issue and enforce permits), or at the offices of the permittee.

You can arrange to either see or obtain copies of the monthly reports or the quarterly summaries of noncompliance generated by the agency receiving them. Public interest groups have recently used these as the basis for lawsuits against the permittees.

Save the Bay, a Rhode Island environmental group, uses the reports to assess regularly the performance of local waste-treatment plants. They base their ratings on average month BOD and TSS concentrations, quantities, and percentage-removal efficiencies. Through annual (now quarterly) surveys, and the publicity that local newspapers have given to their exposure of noncompliance, Save the Bay has brought the issue to the front burner with the public and among state political figures.

The Clean Water Act states that sewage treatment plants were required to achieve secondary levels for BOD and TSS by July 1, 1988. After that date, BOD and TSS discharges were not to exceed a monthly average concentration of 30 milligrams per liter for either and had to have an 85 percent removal efficiency for those pollutants (30/30-85 permits).

Similarly, reports by industrial dischargers can be compared to their permit standards and, if not in compliance, a lawsuit can be filed. The defendant may enter into a consent decree that shows a schedule for compliance and stipulates future penalties for noncompliance. Penalty money generally goes to the U.S. Treasury, but the EPA and the Justice Department can, in a settlement, suggest to the judge that up to half the penalties go toward a closely related environmental mitigation project.

Attorney's fees for the plaintiff can also be recovered from the defendant. Some environmental law firms will work on that basis and will wait to be paid until the settlement is made. Given the nature of the evidence, the risks of losing the litigation can be assessed with fair accuracy.

Direct monitoring by "streamwalking," by a "riverkeeper," or by station sampling and analysis programs has gained a number of proponents in the last few years. The purpose usually is to monitor the state of the environment, i.e., measure dissolved oxygen or clarity, or to identify an illegal discharge.

Finding, sampling, and analyzing polluters that may not have a permit can be critical in uncovering a violation, but such action might be beyond the resources and know-how of the average citizen. You can, however, turn the sample over to a state agency along with your suspicions.

Not all environmentalists agree with citizen-monitoring programs, especially those that require repeated sampling and chemical analyses. They contend that that is the job of the state. Like crime, they say, it is a citizen's duty to report it but not to investigate it.

To find legal help, especially for a lawsuit, discuss your potential case with your nearest law school. They may have an "Environmental Law Clinic" (as does Rutgers University in New Jersey) where you can obtain advice on who might help you with your case. Another source of information is your local bar association which usually has a public interest group.

WIELDING INFLUENCE

Public support is a powerful incentive for political action. Exposing your issue to the light of day may be the most effective way to get something done about it.

Environmental organizations can make or break their image and reputation by the way they comport themselves at hearings, with government officials, and in the media. Too often, through badly chosen rhetoric, environmental activists

stereotype themselves as overly emotional and poorly in-
formed.

Before you meet the press, the public, or officialdom,
choose your spokesperson carefully and prepare your position.
Organize your presentation and back it with facts.

When talking to the press, the North Carolina Coastal
Federation advises, "Make your strongest point first. Add
qualifications later. Accurate, concise statements are the most
effective. Don't swamp a reporter with details . . . It is your
responsibility to make sure reporters understand the context
of any statements you make . . .A reporter cannot be your
ally. They have a professional duty to present both sides of
every issue. If you want your views to be compelling to the
public, you must present a strong case to the reporter."

Save the Bay advises, "Remember, a positive approach
works best. Your organization is 'for' clean water and not
'against' a municipality or its sewage plant administrators and
officials." The same advice applies when dealing with any
agency or organization.

Your choice of tactics, like your choice of issues, will set
the tenor of your organization and determine its effectiveness.
Pick trivial issues and, no matter how popular at the time,
your support will fade away as the world moves on to more
substantial matters. Choose a sanctimonious approach, lapse
into demagoguery, or grow abrasive and you will create resistance to your cause that has little to do with its basic tenets.

Environmental groups that hold what others consider
extreme views often get criticized as "tree huggers" or "eco-
evangelists" who want to bring progress to a halt. Couching
an argument in mysticism, quasi-religious overtones, or highly
charged emotional sentiments reinforces that point of view
among neutral observers. Unless a case is presented with
remarkable skill it's best to follow S. J. Perelman's advice,
"Forget the muses; it's a hard dollar."

THE LAST WORD

Concerns for a cleaner environment are driven mainly by economics or social self-interest—strong motives for the short run but, over time, easily displaced by turns of fortune. Our view of the natural world—its life and substance—must be based on respect for its integrity and guided by a sense of ethics and stewardship.

Aldo Leopold once defined an ecological ethic as "a limitation on freedom of action in the struggle for existence." "All ethics," he said, "rest upon a single premise—that the individual is a member of a community of interdependent parts." Collectively our actions have been antithetical to the inclusion of soil, water, plants, and animals within the community.

We simply do not comprehend the extent to which we have stretched the resiliency of nature, nor do we recognize that nature, bent under these new and strange stresses, is losing its elasticity.

Experience has yet to teach us that neither the private ownership of land nor the use of common water conveys the right to spoil them. Land, water, and wildlife are not artifacts along the course of civilization. They are its roots.

APPENDICES

APPENDIX A

FEDERAL MARINE POLLUTION CONTROL AND RELATED STATUTES

Rivers and Harbors Act of 1899, 33 U.S.C.407

The purpose of the Refuse Act is to keep navigable waters free of obstructions. It prohibits discharge of "any refuse matter of any kind or description whatever other than flowing from streets and sewers and passing therefrom in a liquid state." This law is administered by the U.S. Army Corps of Engineers. Although actively in force, parts of it have been superseded by the Clean Water Act of 1972 (see below).

Water Pollution Control Act of 1948, 1956, 33 U.S.C.1151

These were precursors of present laws on water pollution.

Federal Water Pollution Control Act of 1972 and Amendments, PL 92–500 (The Clean Water Act)

This statute and its amendments (1987 is the latest) are the backbone of U.S. water pollution abatement policy and represent a philosophical break with past laws. The statute has been repeatedly strengthened. EPA is the administrative authority and oversees state implementation. See text for details (pages 00 to 00) and Appendix B for a list of its major provisions.

Oil Pollution Act of 1961, 33 U.S.C. 1001–1015.

This law regulates vessel discharge of oil and construction standards for tankers. It is administered by the Department of Transportation.

National Environmental Policy Act, 42 U.S.C. 4321–4347.

NEPA is a general-purpose statute which requires that environmental impacts be assessed for all federal activities that "significantly affect the quality of the human environment." It created the use of the Environmental Impact Statement (EIS).

Marine Protection, Research, and Sanctuaries Act of 1972, 33 U.S.C. 1401

It is better known as the Ocean Dumping Act. See text for details, pages 00 to 00.

The Coastal Zone Management Act, 16 U.S.C. 1451—1464.

CZMA is directed at the effective management, beneficial use, and protection and development of the coastal zone. See text for details, pages 00 to 00.

Intervention on the High Seas Act, 33 U.S.C. 1371–1387.

This act authorizes the Coast Guard to prevent, mitigate, or eliminate harmful effects of an oil spill when it poses a threat to the U.S. coastline.

Deepwater Port Act of 1974, 33 U.S.C. 1501–1524.

Although its main purpose is to regulate the siting and construction of ports, it provides for the protection of the coastal and marine environment in the course of construction. The law is administered by the Department of Transportation.

Clean Air Act Amendments of 1977, 42 U.S.C. 7401.

This act has resulted in the generation of large amounts of fly ash and flue-gas desulfurization sludge from air-pollution control. Marine disposal has been proposed for these wastes.

Comprehensive Environmental Response, Compensation, and Liability Act Of 1980, 42 U.S.C. 9601.

"Superfund" provides emergency response and cleanup for chemical spills and releases from hazardous waste treatment, storage, and disposal facilities. It has identified large numbers of hazardous waste sites in the coastal zone and has suggested that some wastes generated by remedial action at hazardous waste sites be dumped at sea.

Endangered Species Act of 1973, 16 U.S.C. 1531.

It requires all federal agencies and their permittees to ensure that their actions are unlikely to jeopardize any endangered or threatened species or its habitat.

National Ocean Pollution Planning Act of 1978, 33 U.S.C.1701.

This directs the National Oceanic and Atmospheric Administration to coordinate ocean pollution research and monitoring and to establish priorities for research.

Port and Tanker Safety Act of 1978, 33 U.S.C.1221

It regulates safe handling, loading, storage, and movement of dangerous materials.

Safe Drinking Water Act of 1974, 42 U.S.C.300

This act mandates the development of drinking water standards. Compliance requires upgrading wastewater treatment that generates additional sludge.

Toxic Substances Control Act, 15 U.S.C.2601.

TOSCA regulates the manufacture, processing, distribution, use, and disposal of chemical substances that pose a significant risk to human health.

Low-Level Radioactive Waste Policy Amendments Act of 1986, 42 U.S.C.2021.

This act places the responsibility for disposing of LLWs on the states but it has not worked well and may lead to pressure to go back to sea disposal.

APPENDIX B

MAJOR LEGISLATIVE PROVISIONS
OF THE CLEAN WATER ACT

Section	*Purpose*
104(n)	Directs EPA to establish national estuaries programs to prevent and control pollution; to conduct and promote studies of health effects of estuarine pollution.
104(q)	Establishes a clearinghouse for information on small sewage flows and alternative treatment technologies.
201; 202;203	Specify sewage treatment construction grants, program eligibility, and federal share of cost.
208	Authorizes states and regional authorities to establish planning for point and nonpoint pollution abatement.
301	Directs states to set and periodically upgrade navigable water-quality standards. Sets deadlines for POTW to reach secondary levels.
301(h)	Allows secondary standards waivers for some POTWs.

301(K)   Allows compliance extension for industrial dischargers for development of better effluent-reduction methods.

302   Allows EPA to set additional standards beyond BAT if needed to attain or maintain fishable/swimmable water quality.

303   Requires states to periodically upgrade water-quality standards.

303(e)   Requires states to provide a watershed plan; to control nonpoint source pollution consistent with Sec.208.

304   Requires EPA to establish and periodically revise water-quality criteria to reflect new knowledge about effects and fates of pollutants and to maintain the integrity of navigable waters, ground water, and ocean waters and establish guidelines for effluent limitations.

304(b)   Outlines considerations for effluent limitation guidelines e.g., BPT, BAT, "non-water quality impact," etc.

305(b)   Sets state reporting requirements.

306   Sets new source performance standards.

307   Requires EPA to issue categorical pretreatment standards for new and existing indirect sources; POTWs required to adopt and implement local pretreatment programs. Toxics limits must be set to economically achievable BAT.

308   Requires owners or operators of point sources to maintain records, to sample, to monitor, and to submit data.

309   Gives enforcement powers to states. Sets civil penalties and sanctions and criminal penalties that EPA may impose.

402   Establishes National Pollutant Discharge Elimination System(NPDES) authorizing EPA to issue per-

mits for any pollutant discharged into navigable waters. States given administrative authority.

403   Directs EPA to establish ocean-discharge criteria as guidelines for discharge into territorial seas, contiguous zone, and high seas.

404   Directs secretary of the army to issue permits for dredged or fill material. EPA must establish criteria.

405   Requires EPA to issue sludge use and disposal regulations for POTWs.

504   Grants emergency powers to abate pollutant releases; creates a contingency fund and requires contingency plans for emergencies.

505   Allows citizen civil action in district court against any person in violation of an effluent standard or limitation.

For a thorough review of the act, its history, major features, summary, full legislation, and ancillary amendments, see *The Clean Water Act of 1987* published by the Water Pollution Control Foundation, 601 Wythe St., Alexandria, VA 22314–1944.

APPENDIX C

FEDERAL AND STATE AGENCIES CONCERNED WITH COASTAL WATERS

For a complete listing of federal agencies see:

Conservation Directory, (annual)
National Wildlife Federation
1412 Sixteenth Street,NW
Washington, DC 20036–2266

Coastal Zone Management Programs:

Office of Ocean and Coastal Resources Mgmt.
1845 Connecticut Ave. NW, Suite 700
Washington, DC 20235
(Phone) 202-673-5158

Water Quality:

Environmental Protection Agency
401 M Street SW
Washington, DC 20460

Water Enforcements and Permits, phone: 202-475-8488
Water Regulations and Standards, phone: 202-382-5400
Municipal Pollution Control, phone: 202-382-5850
Marine and Estuarine Protection, phone: 202-475-8580
Wetlands Protection, phone: 202-382-7946

ALABAMA

Dept. of Conservation and Natural Resources
64 N. Union Street
Montgomery, AL 36193
Division of Marine Resources: 205-261-3346

Dept. of Environmental Management
1751 Federal Drive
Montgomery, AL 36130
Phone: 205-271-7700

(CZM)

Office of State Planning and Federal Programs
State Capitol
Montgomery, AL 36130
Phone: 205-284-8706

ALASKA

Dept. of Environmental Conservation
Box 0
Juneau, AK 99811
Phone: 907-465-2600

(CZM)

Coastal Program Coordinator
Office of Management and Budget
Pouch AW–Suite 101
431 North Franklin
Juneau, AK 99811
Phone: 907-465-3562

CALIFORNIA

(CZM)

Federal Programs Manager
California Coastal Commission
631 Howard Street
San Francisco, CA 94105
Phone: 415-543-8555

State Coastal Conservancy
1330 Broadway, Suite 1100
Oakland, CA 94612
Phone: 415-464-1015

CONNECTICUT

(CZM)

CZM Program Director
Department of Environmental Protection
165 Capitol Ave.
Hartford, CT 06106
Phone: 203-566-7404

Council on Environmental Quality
165 Capitol Ave.
Hartford, CT 06106
Phone: 203-566-3510

DELAWARE

(CZM)

Exec. Asst. to the Secretary
Dept. of Natural Resources and Environmental Control
89 Kings Highway, PO Box 1401
Dover, DE 19903
Phone: 302-736-3091

FLORIDA

(CZM)

Coastal Program Manager, OCM
Dept. of Environmental Regulation
2600 Blair Stone Road
Tallahassee, FL 32399
Phone: 904-488-4805

Dept. of Natural Resources
Douglas Bldg.
Tallahassee, FL 32303
Div. of Marine Resources: 904-488-6058

GEORGIA (not participating in CZM)

Dept of Natural Resources
Floyd Towers East
205 Butler Street
Atlanta, GA 30334
Phone: 404-656-3530

HAWAII

(CZM)

CZM Branch Manager
Dept. of Planning and Economic Development
PO Box 2359
Honolulu, HI 96804
Phone: 808-548-4609

LOUISIANA

(CZM)

Coastal Resources Division
Dept. of Natural Resources
PO Box 44396
Capitol Station
Baton Rouge, LA 70804
Phone: 504-342-4500

MAINE

Dept of Marine Resources
State House, Station #21
Augusta, ME 04333
Phone: 207-289-2291

(CZM)

CEIP Coordinator
State Planning Office
184 State Street
Augusta, ME 04330
Phone: 207-289-3261

MARYLAND

(CZM)

Coastal Resources Division
Tawes State Office Bldg.
Annapolis, MD 21401
Phone: 301-269-2784

Dept. of the Environment
201 W. Preston St.
Baltimore, MD 21201
Phone: 301-225-5750

MASSACHUSETTS

(CZM)

Program Manager, CZM
Executive Office of Environmental Affairs
100 Cambridge St.
Boston, MA 02202
Phone: 617-727-9530

Dept of Fisheries, Wildlife, and Environmental
    Law Enforcement
100 Cambridge St.
Boston, MA 02202
Coastal Law Enforcement: 617-727-3190

MISSISSIPPI

(CZM)

Coastal Programs Division
Bureau of Marine Resources
PO Box 959
Long Beach, MS 39560
Phone: 601-864-4602

Dept. of Natural Resources
Bureau of Pollution Control
PO Box 10385
Jackson, MS 39209
Phone: 601-961-5171

NEW HAMPSHIRE

(CZM)

Office of State Planning
2 1/2 Beacon Street
Concord, NH 03301
Phone: 603-271-2155

Dept. of Environmental Services
6 Hazen Drive
Concord, NH 03301
Phone: 603-271-3503

NEW JERSEY

(CZM)

Division of Coastal Resources
Dept. of Environmental Protection
CN 401, Trenton, NJ 08635
Phone: 609-292-2795

NEW YORK

(CZM)

Coastal Program Manager
Div. of Local Government and Community Services
Dept. of State
162 Washington Street
Albany, NY 12231
Phone: 518-474-3643

Dept. of Environmental Conservation
50 Wolf Road
Albany, NY 12233
Public Information, Environmental Quality: 518-457-5400

NORTH CAROLINA

(CZM)

Division of Coastal Management
Dept. of Natural Resources and Community Dev't
PO Box 27687
Raleigh, NC 27611
Phone: 919-733-2293

OREGON

(CZM)

Lead Policy Analyst Manager
Dept. of Land Conservation and Dev't
1175 Court Street, NE
Salem, OR 97310
Phone: 503-378-4926

PENNSYLVANIA

(CZM)

>Division of Coastal Zone Management
>Dept. of Environmental Resources
>PO Box 1467
>Harrisburg, PA 17120
>Phone: 717-787-2814

PUERTO RICO

(CZM)

>Coastal Management Office
>Dept. of Natural Resources
>PO Box 5887
>Puerto de Tierra, PR 00906
>Phone: 809-724-5516

RHODE ISLAND

(CZM)

>Coastal Resources Management Council
>60 Davis Street
>Providence, RI 02908
>Phone: 401-277-2476

>Dept. of Environmental Management
>9 Hayes Street
>Providence, RI 02908
>Div. of Coastal Resources: 401-277-3429

SOUTH CAROLINA

>South Carolina Coastal Council
>4280 Executive Place North, Suite 300
>Charleston, SC 29403
>Phone: 803-744-5830

(CZM)

> Executive Director:
> 1116 NCNB Tower
> Columbia, SC 29201
> Phone: 803-734-1220

TEXAS (not participating in CZM)

> Water Commission
> PO Box 13087
> Capitol Station
> Austin, TX 78711
> Phone: 512-463-8028

VIRGINIA

(CZM)

> Council on the Environment
> Ninth Street Office Bldg. Fl. 9
> Richmond, VA 23219
>
> Marine Resources Commission
> PO Box 756
> 2401 West Ave.
> Newport News, VA 23607
> Phone: 804-247-2200

WASHINGTON

(CZM)

> Shorelands Division
> Dept. of Ecology
> State of Washington (PV-11)
> Olympia, WA 98504
> Phone: 206-459-6777

APPENDIX D

ENVIRONMENTAL ORGANIZATIONS ACTIVE ON COASTAL ISSUES

(R) = regional

American Littoral Society
Sandy Hook
Highlands, NJ 07732
Phone: 201-291-0055

Chesapeake Bay Foundation (R)
162 Prince George Street
Annapolis, MD 21401
Phone: 301-268-8816

Center for Environmental Education
624 9th Street, NW, Suite 500
Washington, DC 20001
Phone: 202-737-3600

Citizens Environmental Coalition (R)
1413 Westheimer Road
Houston, TX 77006
Phone: 713-529-2229

Citizen for a Better Environment (R)
942 Market Place, Suite 505
San Francisco, CA
Phone: 415-788-0690

Clean Ocean Action (R)
PO Box 126
Seabright, NJ 07760
Phone: 201-741-1526

Clean Water Action Project
317 Pennsylvania Ave. SE
Washington, DC 20003
Phone: 202-547-1196

Coalition for the Bight (R)
101 East 15th Street
New York, NY 10003
Phone: 212-460-9250

Coast Alliance
1536 16th Street, NW
Washington, DC 20036
Phone: 202-265-5518

Conservation Council of North Carolina (R)
206 New Bern Place
Raleigh, NC 27601
Phone: 919-755-1329

Committee to Preserve Assateague Island, Inc. (R)
616 Piccadilly Road
Towson, MD 21204
Phone: 301-828-4520

Connecticut Fund for the Environment (R)
32 Grand Street
Hartford, CT 06106
Phone: 203-524-1639

Conservation Law Foundation of New England, Inc. (R)
3 Joy Street
Boston, MA 02108
Phone: 617-742-1540

Environmental Defense Fund
444 Park Ave. South, 9th Fl.
New York, NY 10016
Phone: 212-686-4191

Environmental Health Coalition (R)
PO Box 8426
San Diego, CA 92102
Phone: 619-235-0281

Environmental Policy Institute
218 D Street, SE
Washington, DC 20003
Phone: 202-544-2600

Florida Public Interest Research Group (R)
226 West Pensacola Street
Tallahassee, FL 32301
Phone: 904-224-5304

Friends of the Earth, Northwest (R)
4512 University Way NE
Seattle, WA 98105
Phone: 206-633-1661

Great Lakes United (R)
24 Agassiz Circle
Buffalo, NY 14214
Phone: 716-886-0142

Greenpeace
1611 Connecticut Ave. NW
Washington, DC 20003
Phone: 202-462-1177

Group for the South Fork (R)
PO Box 569
Bridgehampton, NY 11932
Phone: 516-537-1400

Gulf Coast Coalition for Public Health (R)
95 Poinciana Drive, No. 137
Brownsville, TX 78521
Phone: 512-546-0578

Heal the Bay (R)
12234 Pico Blvd.
Los Angeles, CA 90064
Phone: 213-270-4151

Hudson River Fisherman's Assn.
PO Box 312
Cold Spring, NY 10516

Izaak Walton League
1401 Wilson Blvd., Level B,
Arlington, VA 22209
Phone: 703-528-1818

League for Coastal Protection (R)
567 B Avenue
Coronado, CA 92118
Phone: 619-435-4549

Lower James River Association (R)
PO Box 110
Richmond, VA 23201
Phone: 804-730-2898

Maine Audubon (R)
Gilsland Farm
118 Route One
Falmouth, Maine 04105
Phone: 207-781-2330

Mobile Bay Audubon Society (R)
PO Box 9903
Mobile, AL 36609
Phone: 205-666-2476

National Audubon Society
801 Pennsylvania Avenue, SE, Suite 301
Washington, DC 20003
Phone: 202-547-9009

National Coalition for Marine Conservation
1 Post Office Square
Boston, MA 02109
Phone: 617-338-2909

National Wildlife Federation
1412 16th Street, NW
Washington, DC 20007
Phone: 202-797-6800

Natural Resources Defense Council
122 East 42nd Street
New York, NY 10168
Phone: 212-949-0049

New York Audubon Society (R)
8 Wade Road
Latham, NY 12110
Phone: 518-783-8587

North Carolina Coastal Federation (R)
Rt.5, Box 603 (Ocean)
Newport, NC 28570
Phone: 919-393-8185

Oceanic Society
1536 16th Street, NW
Washington, DC 20036
Phone: 202-328-0098

1001 Friends of Oregon (R)
400 Dekum Bldg.
519 Southwest Third Ave.
Portland, OR 97204
Phone: 503-223-4396

Pamlico-Albemarle Study (R)
2108 Dunhill Drive
Raleigh, NC 27608
Phone: 919-833-4859

Pamlico-Tar River Foundation (R)
PO Box 27687
Washington, NC 27889
Phone: 919-946-7211

Puget Sound Alliance (R)
13245 40th Street, NE
Seattle, WA 98125
Phone: 206-343-5865

Save the Bay (R)
434 Smith Street
Providence, RI 02908
Phone: 401-272-3540

Save San Francisco Bay Association (R)
PO Box 925
Berkeley, CA 94701
Phone: 415-849-3044

Sierra Club, Clean Coastal Waters Task Force (R)
1841 North Fuller Ave., Suite 209
Los Angeles, CA 90046
Phone: 213-874-6732

Sierra Club, Delta Chapter (R)
1310 Felicity, B
New Orleans, LA 70130
Phone: 504-586-4751

Sierra Club National Coastal Committee
11194 Douglas Ave.
Marriottsville, MD 21104
Phone: 301-422-5639

South Carolina Wildlife Federation (R)
Box 4186
Arcadian Plaza, Suite B-1
4949 Two Notch Road
Columbia, SC 29240
Phone: 803-786-6419

Southern Environmental Law Center (R)
201 West Main Street, Suite 14
Charlottesville, VA 22901
Phone: 804-977-4090

Texas Environmental Coalition
Box 701
Lake Jackson, TX 77566
Phone: 713-297-6360

Washington Environmental Council (R)
4516 University Way NE
Seattle, WA 98105
Phone: 206-547-2738

Watershed Association of the Delaware River (R)
9A Church Street
Lambertville, NJ 08825
Phone: 609-397-4410

APPENDIX E

ACRONYM GLOSSARY

BAT Best available technology (economically available). EPA's effluent standards for toxic and unconventional pollutants.

BCT Best conventional (pollution control) technology. EPA's standards to be introduced on conventional pollutants after BPT. See sect. 301 and 304 of CWA.

BOD Biochemical oxygen demand. The amount of dissolved oxygen required by microorganisms to oxidize wastes in water or sewage.

BPT Best practicable technology (currently available). EPA's interim standards for municipal and industrial dischargers.

CAA Clean Air Act.

COE U.S. Army Corps of Engineers.

CSO Combined sewer overflow (system). Sewage collection where household and industrial wastes are mixed with storm runoff.

*CWA Clean Water Act. Common name for the Federal Water Pollution Act of 1972, PL-92-500 and subsequent amendments; the latest being PL-100-4, 1987. See Appendix B for major provisions.

*see Appendix A

*CZMA Coastal Zone Management Act, PL 92-583, 1972.

DDT Dichloro diphenyl trichloroethane. Persistent pesticide that biomagnified, causing deaths of apex predators. Banned in US in 1972, as were aldrin and dieldrin in 1974, followed by chlordane and heptachlor in 1976. Still used extensively abroad in third-world countries.

EIS Environmental impact statement. Required of all federal activities that "significantly affect the quality of the human environment." See NEPA in Appendix A.

EPA Environmental Protection Agency.

GAO General Accounting Office.

MARPOL, MARPOL 73/78 International Convention for the Prevention of Pollution from Ships (1973 and Protocols, 1978). International laws whose main concern originally was oil-waste disposal at sea.

MARPOL ANNEX 5 Proposed laws on sewage and garbage disposal at sea not yet ratified by enough nations, including U.S., to become binding.

*MPRSA Marine Protection, Research and Sanctuaries Act, PL 92-532; also known as the Ocean Dumping Act.

NIMBY "Not in my backyard." Refers to public resistance to siting local disposal, collection, or treatment facilities.

NOAA National Oceanic and Atmospheric Administration.

NPDES National Pollutant Discharge Elimination System. Point-source pollution control program, authorized under CWA. Requires permits and sets effluent limits.

OTA Office of Technology Assessment. Federal agency that has performed a series of assessments for Congress on waste issues.

PAH Polycyclic aromatic hydrocarbons. Mainly petroleum derivatives capable of being bioaccumulated or biomagnified. Some are known carcinogens or teratogens.

PCB Polychlorinated biphenyls. A group of over 200 compounds, once used in transformer oil. Remarkably persistent and toxic to marine life. Banned in U.S. in 1979.

POTW Publically-owned treatment works. Federal term for any municipal sewage treatment plant.

ppm Part(s) per million. Roughly equivalent to milligrams per liter.

RCRA Resource, Conservation and Recovery Act. Federal law for the handling, storage, transportation, and disposal of hazardous waste.

TBT Tributyl tin. Organic tin compound used in some anti-fouling paints. Extremely toxic to marine life.

*TOSCA Toxic Substances Control Act. Regulates chemicals that pose significant human health risks.

TSS Total suspended solids. A measure of undissolved waste in water.

FURTHER READING

Clark, J. "Coastal Ecosystem Management", Conservation Foundation, Krieger Publishing Co., Malabar, Fl., 1983. 928pp. Extensive compilation of specific management practices from sewage treatment to tie-downs for trailers.

Kovalic, J. "The Clean Water Act of 1987", Water Pollution Control Federation, Alexandria, Va., 1987. 318pp. History of the law, the law itself, and comments on it. WPCF's address is 601 Wythe Street, Alexandria, Va. 22314

Milleman, B. "And Two If By Sea", Coast Alliance, Washington, DC., 1986. 109pp. Mainly concerned with the Coastal Zone Management Act.

Office of Technology Assessment "Wastes in Marine Environments" OTA-0-334, United States Government Printing Office, Washington, DC, 1987. 312pp. General overview. Good bibliography.

"Saving our Bays, Sounds, and the Great Lakes" Save the Bay, Providence, RI, 1988 83pp. Highlights of a conference of environmental groups.

INDEX

146

2446